Communication
and the
Evolution
of Society

Communication and the Evolution of Society

Jürgen Habermas

Translated and with an Introduction by Thomas McCarthy

Beacon Press : Boston

German Texts: *Sprachpragmatik und Philosophie* and *Zur Rekonstruktion des Historischen Materialismus:* Copyright © 1976 by Suhrkamp Verlag, Frankfurt am Main.

Introduction and English translation: Copyright © 1979 by Beacon Press.

Beacon Press books are published under the auspices of the Unitarian Universalist Association

Published simultaneously in Canada by Fitzhenry & Whiteside Limited, Toronto

"What is Universal Pragmatics?" was originally published as "Was heisst Universalpragmatik" in *Sprachpragmatik und Philosophie,* edited by Karl-Otto Apel, Suhrkamp, Verlag, 1976.

"Moral Development and Ego Identity," "Historical Materialism and the Development of Normative Structures," "Toward a Reconstruction of Historical Materialism," and "Legitimation Problems in the Modern State" were originally published as "Moralentwicklung und Ich-Identitat," "Einleitung: Historischer Materialismus und die Entwicklung normativer Strukturen," "Zur Rekonstruktion des Historischen Materialismus," and "Legitimationsprobleme in modernen Staat," respectively, in *Zur Rekonstruktion des Historischen Materialismus,* by Jürgen Habermas, Suhrkamp Verlag, 1976.

Grateful acknowledgment is made to the following:

American Psychological Association and Jane Loevinger for permission to print the schema "Stages of Ego Development," from the article "The Meaning and Measurement of Ego Development," by Jane Loevinger, *American Psychologist,* Volume 21, No. 3, March 1966.

The Society for Research in Child Development, Inc., for permission to print the schema "Stages of Moral Consciousness," from the article "Conflict and Transition in Adolescent Moral Development," by Elliot Turiel, *Child Development,* 45 (1974).

Academic Press and Lawrence Kohlberg for permission to print the schema "Definition of Moral Stages," from the essay "From Is to Ought," by Lawrence Kohlberg, *Cognitive Development and Epistemology,* Theodore Mischel (ed.), New York, 1971.

Library of Congress Cataloging in Publication Data
Habermas, Jürgen.
    Communication and the evolution of society.
    Includes bibliographical references and index.
    1. Social evolution. 2. Pragmatics. 3. Historical materialism. 4. State, The. 5. Social sciences—Philosophy. I. Title.
HM106.H313   301.14   77-88324
ISBN 0-8070-1521-1   ISBN 0-8070-1513-X pbk.

Contents

Translator's Introduction

Some twenty years ago Jürgen Habermas introduced his idea of a critical social theory that would be empirical and scientific without being reducible to empirical-analytic science, philosophical in the sense of critique but not of presuppositionless "first philosophy," historical without being historicist, and practical in the sense of being oriented to an emancipatory political practice but not to technological-administrative control.[1] Although these general features are still recognizable in his mature views on critical theory, the original conception has undergone considerable development. The essays translated in this volume provide an overview of the theoretical program that has emerged. Before sketching its main lines it might be well, by way of introduction, to review briefly Habermas' earlier discussions of social theory; for in these a number of important ideas that have since receded into the background or altogether disappeared from view are still clearly visible.

I

A recurring theme of Habermas' writings in the late fifties and early sixties was that critique must somehow be located "between philosophy and science."[2] In his account of the transition from the classical doctrine of politics to modern political science, Habermas noted a decisive shift in the conceptions of theory and practice and their interrelation.[3] For Aristotle politics was continuous with ethics, the doctrine of the good and just life. As such it referred to the sphere of human action, *praxis,* and was directed to achieving and maintaining an order of virtuous conduct among the citizens of the *polis.* The practical intention of politics, as well as the nature of its subject matter, determined its cognitive status: Politics could not assume the form of a rigorous science, of *episteme* but had to rest content with establishing rules of a more-or-less and in-most-cases character. The capacity thereby cultivated, and the keystone of the virtuous character, was *phronesis,* a prudent understanding of variable situations with a view to what was to be done.

With the rise of modern science the classical conception of politics was drastically altered. Theory came to mean the logically integrated systems of quantitatively expressed, lawlike statements characteristic of the most advanced sciences. Given a description of the relevant initial conditions, such theories could be used (within certain limits) to predict future states of a system; providing the relevant factors were manipulable, they could also be used to produce desired states of affairs. Adopting this ideal of knowledge for politics, Hobbes early outlined a program that took human behavior as the material for a science of man, society, and the state. On the basis of a correct understanding of the laws of human nature it would be possible to establish once and for all the conditions for a proper ordering of human life. The classical instruction in leading a good and just life, the formation of virtuous character, and the cultivation of practical prudence would be replaced by the application of a scientifically grounded social theory, by the production of the conditions that would lead to the desired behavior according to the laws of nature. In this way the sphere of the practical was absorbed into the sphere of the technical; the practical problem of the virtuous life of the citizens of the *polis* was transformed into the technical-administrative problem of regulating social intercourse so as to ensure the order and well-being of the citizens of the state.

In Habermas' view the principal loss incurred in this transition was the replacement of a direct access to practice with a purely technological understanding of the theory-practice relationship; the principal gain was the introduction of scientific rigor into the study of society. Accordingly, the outstanding task for a post-positivist methodology of social inquiry was somehow to combine philosophical and practical moments with the methodological rigor, which was "the irreversible achievement of modern science." [4] Of course, the type of practical philosophy Habermas himself had in mind was not the classical Greek but that which developed in the movement of German thought from Kant through Marx; and the type of combination he envisaged was summed up in the phrase: "empirical philosophy of history with a practical (political) intent."

The presence of the term *philosophy* in this characterization of critical theory did not signal a basic disagreement with Marx's dictum that the demands and results of philosophy could be preserved only through "the negation of previous philosophy, of philosophy as philosophy." Habermas was not using the term in its traditional sense as a presuppositionless mode of thought that provided its own foundations. With Marx he regarded philosophy as belonging to the world on which it reflected and as having to return to it; the ideals inherent in philosophy—truth and reason, freedom and justice—could not be realized by thought itself. The philosophy of history, in particular, was marred by a failure to realize this. Pretending to a contemplative view of the whole of history, prospective as well as retrospective, it claimed to reveal its meaning, often in terms of a necessary progress toward some metaphysically guaranteed goal ascribed to God or Nature, Reason or Spirit.

As Habermas interpreted him, the young Marx rejected this construction. For him the movement of history was not at all a matter of metaphysical necessity; it was contingent in regard to both the empirical conditions of change and the practical engagement of social agents. The meaning of history, its goal, was not a subject for metaphysical hypostatization but for practical projection; it was a meaning that men, in the knowledge of objective conditions, could seek to give it with will and consciousness. The exaggerated epistemic claims of the traditional philosophy of history derived in part from ignoring the essentially practical nature of its prospective dimension. The projected future (which conferred meaning on the past) was not a product of contemplation or of scientific prediction but of a situationally engaged practical reason.

The meaning of the actual historical process is revealed to the extent that we grasp a meaning, derived from "practical reason," of what should be and what should be otherwise . . . and theoretically examine the presuppositions of its practical realization . . . We must interpret the actual course and the social forces of the present from the point of view of the realization of that meaning.[5]

Thus Habermas already found in the young Marx many of the necessary correctives to the excesses of traditional philosophy. But Marx, in his desire to distinguish himself from the "merely philosophic" critique of the left Hegelians, subsequently ascribed to his own views the features of a strictly empirical theory of society; and later, in the hands of his "orthodox" followers, Marxism seemed to provide a purely theoretical guarantee of the outcome of history; the importance of critical self-reflection and enlightened political practice receded behind the solid, objective necessity of inexorable laws of history. The spectacle of this retrogression was one of the motivating factors behind the Frankfurt School's renewal of the philosophical dimension of Marxism; and it was behind Habermas' concern to demarcate critical social theory from strictly empirical-analytic science as clearly as Marx had from philosophy—to locate it "between philosophy and science."

While the essays of the late fifties and early sixties introduced the idea of comprehending society as a historically developing whole for the sake of enlightening practical consciousness, building a collective political will, and rationally guiding practice, they provided as yet no detailed articulation of the logic, methodology, or structure of this type of theory. The first attempts to do so appeared in the later sixties, principally in *Zur Logik der Sozialwissenschaften* (1967) and *Knowledge and Human Interests* (1968).[6] Although these studies were still labelled "propaedeutic" by Habermas, they did contain extended discussions

of the methodological issues surrounding social inquiry in general
and critical theory in particular. One of his principal targets in
both books was the neopositivist thesis of the unity of scientific
method, the thesis, in particular, that the logic of scientific inquiry
in the human sciences is basically the same as that in the natural
sciences. In *Zur Logik der Sozialwissenschaften* the main line of
attack runs through a consideration of the nature and role of
Verstehen, or interpretive understanding, in social inquiry. Ex-
aming various *verstehenden* approaches to society—neo-Kantian
and Weberian, social interactionist, phenomenological and ethno-
methodological, linguistic and hermeneutic—Habermas argues
that access to a symbolically structured object domain calls for
procedures that are logically distinct from those developed in the
natural sciences, procedures designed to grasp the "meaning"
that is constitutive of social reality. Social action depends on the
agent's "definition of the situation," and this is not solely a matter
of subjective motivations. The meanings to which social action is
oriented are primarily intersubjective meanings constitutive of
the sociocultural matrix in which individuals find themselves and
act: inherited values and world views, institutionalized roles and
social norms, and so on. Any methodology that systematically
neglects the interpretive schemata through which social action is
itself mediated, that pursues the tasks of concept and theory
formation in abstraction from the prior categorical formation of
social reality, is doomed to failure. Sociological concepts are, in
Alfred Schutz's words, "second-level constructs"; the "first-level
constructs" are those through which social actors have already
prestructured the social world prior to its scientific investigation.
Understanding the latter is a necessary point of departure for
constructing the former.

   While arguing this point Habermas was careful, at the same
time, to distance himself from the view that interpretive understanding could be the sole methodological basis of social inquiry.
In his lengthy discussion of Gadamer's philosophical hermeneutics,
which he took to be the most developed form of this view, he
pointed out different aspects of social reality that called for modes
of inquiry going beyond the merely interpretive.[7] For one thing,
the reduction of social research to the explication of meaning

rests on an unwarranted sublimation of social processes entirely into subjectively intended and/or culturally transmitted meanings. If, however, these meanings are viewed in relation to the social, political, and economic conditions of life, it becomes evident that they can conceal and distort as well as reveal and express these conditions. Thus an adequate social methodology would have to integrate interpretive understanding with critique of ideology. Of course, this requires a system of reference that goes beyond subjective intentions and cultural tradition, one that systematically takes into account the objective framework of social action and the empirical conditions under which traditions historically change. Developments in the economic and political spheres, for example, can overturn accepted patterns of interpretation. And such developments are not as a rule simply the results of new ways of looking at things; rather they themselves bring about a restructuring of world views. Thus an adequate social methodology would have to integrate interpretive understanding and critique of ideology with an historically oriented analysis of social systems.

To specify desiderata in this way is obviously only a first step on the way to a fully developed critical social theory. In both *Zur Logik der Sozialwissenschaften* and *Knowledge and Human Interests* Habermas did go on to offer a number of suggestions on the direction in which further steps might lead. In the latter work, he used Freudian psychoanalysis as a "tangible example" of critical theory in order to derive from its analysis a number of general methodological clues.[8] Interpreting Freud's work as a theory of systematically distorted communication, he pointed out the ways in which it went beyond a purely *verstehenden* explication of meaning. In contrast to normal hermeneutics, psychoanalytic interpretation deals with "texts" that both express and conceal their "author's" self-deceptions. The "depth hermeneutics" that Freud developed to deal with this "internal foreign territory" relies on theoretical assumptions that are only partly explicit in his own work. Their full and consistent development would require a general theory of normal (undistorted) communication, a developmental account of the acquisition of the competence to communicate, as well as an account of the conditions under which systematic distortions in communication arise. It is on this last point that Freud has most to offer; he provides us with a general interpretation of early-childhood patterns of interaction, coordinated with a phase-specific model of personality formation. This "general interpretation" or "interpretative schema" has the form of a "systematically generalized history" of psychodynamic development. Its methodological peculiarities provide clues as to what is distinctive about critical theory. For one thing, the application of such an interpretive schema has an inexpungible hermeneutic component. Its concepts are schematic or type concepts that have to be translated into individuated situations; it is applied in constructing histories in which subjects can recognize themselves and their world. In contrast to ordinary philological hermeneutics, however, this reconstruction of individual life histories requires a peculiar combination of interpretive understanding and causal explanation. "We cannot 'understand' the 'what'—the semantic content of the systematically distorted expression—without at the same time 'explaining' the 'why'— the origin of the systematic distortion itself." [9] The explanatory hypotheses refer not to the "causality of nature" but, so to speak, to the "causality of fate," that is, to the workings of repressed motives and other "symbolic contents." The postulated causal connections do not represent an invariance of natural laws but an invariance of life history that operates through "the symbolic means of the mind" and can thus be analytically dissolved.

Other methodological peculiarities of Freud's general theory of psychodynamic development concern the type of corroboration appropriate to a systematically generalized history of this type. The assumptions it contains—about interaction patterns between the child and primary reference persons, about corresponding conflicts and forms of coping with conflict, about the personality structures that result, and so on—serve as a "narrative foil" for the reconstruction of individual life histories. They are developed as the result of numerous and repeated clinical experiences and are correspondingly subject to empirical corroboration. But this corroboration is of a distinctive sort; the physician's attempt to combine the fragmentary information obtained in the analytic dialogue and to offer a hypothetical reconstruction of the patient's life history essentially anticipates the latter's own reflective appropriation of this story. The corroboration of a general interpretation thus ultimately relies on the successful continuation of processes of self-formation: "only the context of the self-formative process as a whole has confirming and falsifying power." [10]

The relevance of this notion of a "systematically" or "theoretically generalized history" for the critical theory of *society* was suggested in Habermas' discussion of Parsons in *Zur Logik der Sozialwissenschaften*.[11] Structural-functionalism interested him as an attempt to integrate action-theoretic and systems-theoretic perspectives. Parsons does not ignore the meaningfulness of social action; but he does not limit its significance to what is intended by social agents or articulated in the cultural tradition. The social system is conceived as a functional complex of institutions within which cultural patterns or values are made binding for action, that is, are incorporated into binding social norms and institutionalized values. In this framework it is possible to investigate empirical connections between social norms that go beyond the subjective intentions of those acting under the norms. The significance of the objective connections within the system of social roles is latent; to grasp it we must discover the functions that specific elements fulfill for the self-maintenance of the social system.

Habermas' criticisms of this approach centered around its subordination of the hermeneutic and critical moments of social inquiry to the requirements of empirical-analytic science. Parsons short-circuits the hermeneutic dimension by, for example, adopting the simplifying assumption of a universal value schema; all value systems are constructed from the same set of basic value orientations (pattern variables) fundamental to all social action. But both the universality and the completeness of his table of categories can be questioned; upon closer analysis it becomes evident that the four pairs of alternative value orientations are tailored to an analysis of one historical process, the transformation from traditional to modern society. There is a preunderstanding of the historical situation incorporated into the very formulation of these basic concepts. If the historically situated character of functional analysis is to be taken into account, the problems and methods of historical-hermeneutic reflection become unavoidable.

The critical dimension of social inquiry is also cut short in structural-functional analysis, for it does not permit a systematic separation of the utopian, purposive-rational, and ideological contents of value systems. According to Parsons, cultural values are made binding for social action in institutions; the latter integrate "value orientations" and "motivational forces," thus securing the normative validity of social roles. Habermas found this construction overly harmonistic.

In the framework of action theory, motives for action are harmonized with institutional values, that is, with the intersubjectively valid meaning of normatively binding behavioral expectations. Nonintegrated motive forces that find no licensed opportunity for satisfaction in the role system are not analytically grasped. We may assume, however, that these repressed needs, which are not absorbed into social roles, transformed into motivations, and sanctioned, nevertheless have their interpretations. Either these interpretations "overshoot" the existing order and, as utopian anticipations, signify a not-yet-successful group identity; or, transformed into ideologies, they serve projective substitute gratification as well as the justification of repressing authorities ... In relation to such criteria, a state of equilibrium would be determined according to whether the system of domination in a society realized the utopian elements and dissolved the ideological contents to the extent that the level of productive forces and technical progress made objectively possible. Of course, society can then no longer be conceived as a system of self-preservation ... Rather, the meaning, in relation to which the functionality of social processes is measured, is now linked to the idea of a communication free from domination.[12]

As these last lines indicate, the incorporation of historico-hermeneutic and critical moments into the analysis of social systems bursts the functionalist framework, at least insofar as the latter is understood on the model of biology. The validity of functional analysis presupposes (among other things) that it is possible to specify empirically the boundaries of the system in question, the goal state the system tends to achieve and maintain, the functional requirements for self-maintenance, and the alternative processes through which they can be met. This is the case above all in biology; an organism is easily demarcated from its environment and the state in which it maintains itself can be characterized in terms of necessary processes with specifiable tolerances. The same cannot be said for social systems. In the course of history not only the elements but the boundaries and the goal states of societies undergo change; consequently, their identity becomes blurred. A given modification might be regarded either as a learning process and regeneration of the original system or a process of dissolution and transformation into a new system. There is apparently no way to determine which description is correct independently of the interpretations of members of the system.[13]

Habermas concluded that if social systems analysis incorporated the historico-hermeneutic and critical dimensions as suggested, it could no longer be understood as a form of strictly empirical-analytic science; it would have to be transformed into a historically oriented theory of society with a practical intent. The form such a theory would take was that of a "theoretically generalized history" or "general interpretation" which reflectively grasped the formative process of society as a whole, reconstructing the contemporary situation with a view not only to its past but to its anticipated future. It would be a critical theory of society.

II

On Habermas' own account the methodological views advanced in *Zur Logik der Sozialwissenschaften* and *Knowledge and Human Interests* do not represent a final statement of his idea of a critical social theory.[14] He sees them rather as guideposts on his way to formulating a systematic conception; this latter task has been the focus of his work for the past decade. The essays collected in this volume provide an overview of the results. As Habermas repeatedly reminds us, they are not "results" in the sense of "finished products"; his conception of critical theory is presented rather as a "research program." While he is concerned to argue its validity, he is aware of its hypothetical status, aware that a program of this magnitude requires considerable development before its fruitfulness—theoretical and practical—can be adequately judged. Nevertheless, its main outlines have taken on a definite shape in recent years. It might best be described as a three-tiered research program. The ground level consists of a general theory of communication—as Habermas calls it, a universal pragmatics—at the next level this theory serves as the foundation for a general theory of socialization in the form of a theory of the acquisition of communicative competence; finally, at the highest level, which builds on those below it, Habermas sketches a theory of social evolution which he views as a reconstruction of historical materialism. In the remainder of this introduction, I shall make a few general remarks about each of these subprograms and about Habermas' application of the ideas developed in them to the analysis of contemporary society.

1. As mentioned above, one of the conclusions of Habermas' examination of psychoanalysis was that, as a theory and therapy of systematically distorted communication, it necessarily presupposed a general theory of (nondistorted) communication. This is only a particular instance of a more general conclusion he had reached earlier: that the normative-theoretical foundations of critical theory would have to be sought in that distinctive and pervasive medium of life at the human level, viz. language. In his inaugural lecture of June 1965 at Frankfurt University, he had declared: "What raises us out of nature is the only thing whose nature we can know: language. Through its structure autonomy and responsibility are posited for us. Our first sentence expresses unequivocally the intention of universal and unconstrained consensus. Autonomy and responsibility together (*Mündigkeit*) comprise the only idea we possess a priori in the sense of the philosophical tradition." [15] Of course at that time this was little more than a declaration—that the normative-theoretical foundations of critical theory were badly in need of renewal, that neither dialectical materialism nor a retreat to pure philosophy was adequate to this task, that earlier attempts by the members of the Frankfurt School to articulate and ground a conception of rationality that essentially transcended the narrow confines of "instrumental" thought had not in the end succeeded, and that the solution was to be found in a theory of language.

It is only with the formulation of the basic ideas of his communication theory that this declaration has assumed the more definite form of a research program. The first essay translated (and somewhat revised) for this volume, "What is Universal Pragmatics?," provides the best available statement of the strategy and structure of that program.[16] The central idea is introduced by way of contrast to the usual restriction of rational reconstruction to the syntactic and semantic features of language in abstraction from its pragmatic dimension, which is normally brought in subsequently as a domain for empirical (rather than logical or purely linguistic) analysis. The idea of a universal pragmatics rests on the contention that not only phonetic, syntactic, and semantic features of *sentences,* but also certain pragmatic features of *utterances,* not only language but speech, not only linguistic competence but communicative competence, admit of rational reconstruction in universal terms. Habermas is arguing then "that communicative competence has as universal a core as linguistic competence. A general theory of speech action would thus describe that fundamental system of rules that adult subjects master to the extent that they can fulfill the conditions for a happy employment of sentences in utterances, no matter to which individual languages the sentences may belong and in which accidental contexts the utterances may be embedded." The competence of the ideal speaker must be regarded as including not only the ability to produce and understand grammatical sentences but also the ability to establish and understand those modes of communication and connections with the external world through which speech becomes possible. Pragmatic rules for situating sentences in speech actions concern the relations to reality that accrue to a grammatically well-formed sentence in being uttered. The act of utterance situates the sentence in relation to external reality ("the" world of objects and events about which one can make true or false statements), to internal reality (the speaker's "own" world of intentional experiences that can be expressed truthfully/sincerely or untruthfully/insincerely), and to the normative reality of society ("our" social life-world of shared values and norms, roles and rules, that an act can fit or fail to fit, and that are themselves either right—legitimate, justifiable—or wrong). Regarded from this pragmatic point of view, it becomes clear that speech necessarily (even if often only implicitly) involves the raising, recognizing, and redeeming of "validity claims." In addition to the (implicit) claim that what he utters is comprehensible, the speaker also claims that what he states is true (or if no statement is made, that the existential presuppositions of his utterance's propositional content are fulfilled); that his manifest expression of intentions is truthful or sincere; and that his utterance (his speech *act*) is itself right or appropriate in relation to a recognized normative context (or that the normative context it fits is itself legitimate). The claims to truth, truthfulness, and rightness place the speaker's utterance in relation to extralinguistic orders of reality; the universal-pragmatic infrastructure of speech consists of general rules for arranging the elements of speech situations within a coordinate system formed by "the" external world, one's "own" internal world, and "our" shared social life-world. It is obvious that a fully developed universal pragmatics would provide a unifying framework for a variety of theoretical endeavors usually assigned to disparate and only occasionally related disciplines— from the theory of knowledge to the theory of social action.

2. It was a characteristic tenet of the early Frankfurt School that basic psychological concepts had to be integrated with basic socioeconomic concepts because the perspectives of an autonomous ego and an emancipated society were essentially interdependent. In this way, critical theory was linked to a concept of the autonomous self that was, on the one hand, inherited from German Idealism but was, on the other hand, detached from idealist presuppositions in the framework of psychoanalysis. Habermas too starts from the interdependence of personality structures and social structures, of forms of identity and forms of social integration; but the socio-psychological framework he deploys involves much more than a readaptation of psychoanalysis. It is an integrated model of ego (or self-) development that draws on developmental studies in a number of areas, ranging from psycholinguistics and cognitive psychology (including studies of moral consciousness) to social interactionism and psychoanalysis (in-

cluding analytic ego psychology).[17] The task, as he sees it, is to work out a unified framework in which the different dimensions of human development are not only analytically distinguished but in which their interconnections are also systematically taken into account. Beyond this, the empirical mechanisms and boundary conditions of development have to be specified. This is clearly an immense task, and Habermas is still in the process of working out an adequate research program. The general (and tentative) outlines of his approach are nevertheless clear. He adopts a competence-development approach to the foundations of social action theory; the basic task here is the rational reconstruction of universal, "species-wide," competences and the demonstration that each of them is acquired in an irreversible series of distinct and increasingly complex stages that can be hierarchically ordered in a developmental logic. The dimensions in which he pursues this task correspond to the universal-pragmatic classification of validity claims, that is, to the four basic dimensions in which communication can succeed or fail: comprehensibility, truth, rightness, and truthfulness. Each of these specifies not only an aspect of rationality, but a "region" of reality—language, external nature, society, internal nature—in relation to which the subject can become increasingly autonomous. Thus ontogenesis may be construed as an interdependent process of linguistic, cognitive, interactive, and ego (or self-) development.

Only the first three of these can be regarded as particular lines of development; the ontogenesis of the ego is not a development separable from the others but a process that runs complementary to them: the ego develops in and through the integration of "internal nature" into the structures of language, thought, and action. Of course, the acquisition of universal competences represents only one, the structural, side of identity formation; the other side is affect and motive formation. Unless the subject is able to interpret his needs adequately in these structures, development may be pathologically deformed. Thus a general theory of ego development would have to integrate an account of the interdependent development of cognitive, linguistic, and interactive development with an account of affective and motivational development.

The second essay translated for this volume, "Moral Development and Ego Identity," focuses on one strand of this complex: the development of moral consciousness. Using Kohlberg's hierarchical schema for the ability to make moral judgments, Habermas places it in a larger action-theoretic framework by coordinating the stages of this ability with stages in the development of interactive competence: "I shall proceed on the assumption that 'moral consciousness' signifies the ability to make use of interactive competence for *consciously* dealing with morally relevant conflicts." He then goes on (in part IV of the essay) to consider the motivational (as distinguished from the structural or "cognitive") side of moral consciousness, that is, the psychodynamics of developmental processes (formation of superego, defense mechanisms). This perspective makes it possible to comprehend the frequent discrepancies between moral judgment and moral action. The essay as a whole provides an example of how new perspectives are opened by viewing the separate domains of developmental studies in an integrated framework with both structural and affective-motivational aspects.

3. The third and fourth essays delineate the properly sociological level of Habermas' program: the theory of social evolution. He understands this as a reconstruction of historical materialism, which turns on the thesis that developments in the sphere of social integration have their own logic: "I am convinced that normative structures do not simply follow the path of development of reproductive processes . . . but have an internal history." This is obviously the fruit of his long-standing insistence that *praxis* cannot be reduced to *techne,* nor rationality to purposive or instrumental rationality, that rationalization processes in the sphere of communicative action or interaction are neither identical with nor an immediate consequence of rationalization processes in the sphere of productive forces. In working out the logic of development of normative structures, Habermas' strategy is to employ structural comparisons with the developmental logic worked out for ontogenetic processes in the framework of his theory of communicative competence. This is, of course, a new version of an old strategy, and there is no lack of historical example for the pitfalls that attend drawing parallels between in-

dividual and social development. Habermas is aware of these pitfalls but argues that under certain restrictions one can indeed find "homologous structures of consciousness" in the histories of the individual and the species.

In "Historical Materialism and the Development of Normative Structures" he suggests three domains of comparison: rationality structures in ego development and in the evolution of world views; the development of ego and of group (or collective) identities; the development of moral consciousness and the evolution of moral and religious representations. After sketching briefly the homologous patterns he finds in the first two areas, he turns in the fourth essay, "Towards a Reconstruction of Historical Materialism," to a more detailed examination of the development of law and morality. The explanatory schema advanced there—a combination of action-theoretic (in the competence-development sense) and systems-theoretic motifs—makes it clear that he is not proposing to read human history as an internal unfolding of Spirit. There is an explicit distinction drawn between the logic of development of normative structures and the dynamics of this development. The former merely circumscribes the logical sphere in which increasingly complex structural formations can take shape; whether new structures arise, and if so, when, depends on contingent boundary conditions and empirical learning processes.

The following are the principal elements of the schema: Social evolution is conceived a bidimensional learning process (cognitive/technical and moral/practical), the stages of which can be described structurally and ordered in a developmental logic. The emphasis is not on the institutionalization of particular contents (e.g., values; cf. Parsons), but on the "institutional embodiment of structures of rationality," which makes learning at new levels possible, that is, on learning applied to the structural conditions of learning. In one sense it is only socialized individuals who learn; but the learning ability of individuals provides a "resource" that can be drawn upon in the formation of new social structures. The results of learning processes find their way into the cultural tradition; they comprise a kind of cognitive potential that can be drawn upon in social movements when unsolvable system problems require a transformation of the basic forms of social integration. Whether and how problems arise that overload the structurally limited adaptive capacity of a society, is contingent; whether the necessary, but not yet institutionalized, structures of rationality (technical and practical) are available, whether social movements arise to meet the challenge by drawing on this potential, whether they succeed in institutionalizing new forms of social integration, and whether these institutions can be stabilized, are also dependent on contingent circumstances. Nevertheless the structural descriptions of the different stages of development can be ordered in a developmental logic, that is, in a hierarchical sequence of increasingly complex and encompassing forms of rationality.

4. Critical theory does not exhaust itself in the construction of a theory of social evolution (the reconstruction of historical materialism); its ultimate aim remains an historically oriented analysis of contemporary society with a practical intent (a reconstruction of the critique of capitalist society). By comparison to the retrospective explanation of past developments, the projective analysis of contemporary society has an immediately practical reference.

Evolutionary statements about contemporary social formations have an immediately practical reference insofar as they serve to diagnose developmental problems. The restriction to retrospective explanations of historical material is dropped in favor of a *retrospective projected from the perspectives of action;* the diagnostician of the present adopts the fictive standpoint of an evolutionary explanation of a future past . . . As a rule, Marxist explanations of developed capitalism also share this asymmetric position of the theoretician who analyzes developmental problems of the contemporary social system with a view to structual possibilities that are not yet (and perhaps never will be) institutionalized. It can be seen from this that the application of evolutionary theories to the present makes sense only in the framework of a discursive formation of the will, that is, in a practical argumentation dealing with reasons why specific actors in specific situations ought to choose specific strategies of action over others.[18]

Habermas' principal contribution to the analysis of contemporary capitalism is to be found in *Legitimation Crisis*.[19] The last essay translated for the present volume, "Legitimation Problems in the Modern State," provides an elaboration on the argument advanced there, some responses to critics, and several important conceptual clarifications. The evolution-theoretic background to the argument can be seen in the concept of *levels of justification*— formal conditions for the acceptability of different *kinds* of grounds or reasons, for the efficacy of different *types* of legitimation, for their power to produce consensus and shape motives. The crux of the argument is that legitimation problems arise in developed capitalist societies as the result of a fundamental conflict built into their very structure, a conflict between the social welfare responsibilities of mass democracies and the functional conditions of the capitalist economy. The state is forced to deal with the dysfunctional side effects of the economic process under a number of restrictive conditions—balancing a policy of economic stability against a policy of social reform in a world economy that increasingly limits the individual state's latitude for action and without being able effectively to control social integration or to "plan ideology." To the extent that it fails to keep these side effects within acceptable bounds, manifestations of delegitimation appear—for example, a sharpened struggle over distribution, economic instability, the breakdown of reform politics, and even the disintegration of motivational patterns essential to capitalist society and the spread of dysfunctional patterns. As those familiar with the argument of *Legitimation Crisis* will recall, it is this last level of delegitimation that Habermas regards as fundamental. If the form of life reflected in such system-conforming rewards as money, free time, and security can no longer be convincingly legitimated, "the 'pursuit of happiness' might one day mean something different—for example, not accumulating material objects of which one disposes privately, but bringing about social relations in which mutuality predominates and satisfaction does not mean the triumph of one over the repressed needs of the other."

I would like to express my gratitude to the Alexander von Humboldt Foundation for a grant in the spring of 1978 that enabled me to complete this translation.

Thomas McCarthy
Boston University

1   What Is Universal Pragmatics?*

I

The task of universal pragmatics is to identify and reconstruct universal conditions of possible understanding [*Verständigung*].[1] In other contexts one also speaks of "general presuppositions of communication," but I prefer to speak of general presuppositions of communicative action because I take the type of action aimed at reaching understanding to be fundamental. Thus I start from the assumption (without undertaking to demonstrate it here) that other forms of social action—for example, conflict, competition, strategic action in general—are derivatives of action oriented to reaching understanding [*verständigungsorientiert*]. Furthermore, as language is the specific medium of understanding at the sociocultural stage of evolution, I want to go a step further and single out explicit speech actions from other forms of communicative action. I shall ignore nonverbalized actions and bodily expressions.[2]

The Validity Basis of Speech

Karl-Otto Apel proposes the following formulation in regard to the general presuppositions of consensual speech actions: to

* I would like to thank E. Tugendhat and G. Grewendorf for their helpful criticisms of a first draft of this paper. They will have their disagreements with this revised version as well. J. H.

identify such'presuppositions we must, he thinks, leave the perspective of the observer of behavioral facts and call to mind
"what we must necessarily always already presuppose in regard
to ourselves and others as normative conditions of the possibility
of understanding; and in this sense, what we must necessarily
always already have accepted." ³ Apel uses the aprioristic perfect
[*immer schon*: always already] and adds the mode of necessity
to express the transcendental constraint to which we, as speakers,
are subject as soon as we perform or understand or respond to a
speech act. In or after the performance of this act, we can become
aware that we have involuntarily made certain assumptions, which
Apel calls "normative conditions of the possibility of understanding." The addition "normative" may give rise to misunderstanding. Indeed one can say that the general and unavoidable—in this
sense transcendental—conditions of possible understanding have
a normative content when one has in mind not only the binding
character of norms of action or even the binding character of
rules in general, but the validity basis of speech across its entire
spectrum. To begin, I want to indicate briefly what I mean by
"the validity basis of speech."

I shall develop the thesis that anyone acting communicatively
must, in performing any speech action, raise universal validity
claims and suppose that they can be vindicated [or redeemed:
einlösen]. Insofar as he wants to participate in a process of reaching understanding, he cannot avoid raising the following—and
indeed precisely the following—validity claims. He claims to be:

   a. *Uttering* something understandably;
   b. Giving [the hearer] *something* to understand;
   c. Making *himself* thereby understandable; and
   d. Coming to an understanding *with another person.*

The speaker must choose a comprehensible [*verständlich*] expression so that speaker and hearer can understand one another. The
speaker must have the intention of communicating a true [*wahr*]
proposition (or a propositional content, the existential presuppositions of which are satisfied) so that the hearer can share the
knowledge of the speaker. The speaker must want to express his
intentions truthfully [*wahrhaftig*] so that the hearer can believe

the utterance of the speaker (can trust him). Finally, the speaker must choose an utterance that is right [*richtig*] so that the hearer can accept the utterance and speaker and hearer can agree with one another in the utterance with respect to a recognized normative background. Moreover, communicative action can continue undisturbed only as long as participants suppose that the validity claims they reciprocally raise are justified.

The goal of coming to an understanding [*Verständigung*] is to bring about an agreement [*Einverständnis*] that terminates in the intersubjective mutuality of reciprocal understanding, shared knowledge, mutual trust, and accord with one another. Agreement is based on recognition of the corresponding validity claims of comprehensibility, truth, truthfulness, and rightness. We can see that the word *understanding* is ambiguous. In its minimal meaning it indicates that two subjects understand a linguistic expression in the same way; its maximal meaning is that between the two there exists an accord concerning the rightness of an utterance in relation to a mutually recognized normative background. In addition, two participants in communication can come to an understanding about something in the world, and they can make their intentions understandable to one another.

If full agreement, embracing all four of these components, were a normal state of linguistic communication, it would not be necessary to analyze the process of understanding from the dynamic perspective of *bringing about* an agreement. The typical states are in the gray areas in between: on the one hand, incomprehension and misunderstanding, intentional and involuntary untruthfulness, concealed and open discord; and, on the other hand, pre-existing or achieved consensus. Coming to an understanding is the process of bringing about an agreement on the presupposed basis of validity claims that can be mutually recognized. In everyday life we start from a background consensus pertaining to those interpretations taken for granted among participants. As soon as this consensus is shaken, and the presupposition that certain validity claims are satisfied (or could be vindicated) is suspended, the task of mutual interpretation is to achieve a new definition of the situation which all participants can share. If their attempt fails, communicative action cannot be continued. One is then basically confronted with the alternatives of switching to strategic action, breaking off communication altogether, or recommencing action oriented to reaching understanding at a different level, the level of argumentative speech (for purposes of discursively examining the problematic validity claims, which are now regarded as hypothetical). In what follows, I shall take into consideration only consensual speech actions, leaving aside both discourse and strategic action.

In communicative action participants presuppose that they know what mutual recognition of reciprocally raised validity claims means. If in addition they can rely on a shared definition of the situation and thereupon act consensually, the background consensus includes the following:

a. Speaker and hearer know implicitly that each of them has to raise the aforementioned validity claims if there is to be communication at all (in the sense of action oriented to reaching understanding).

b. Both suppose that they actually do satisfy these presuppositions of communication, that is, that they could justify their validity claims.

c. Thus there is a common conviction that any validity claims raised are either—as in the case of the comprehensibility of the sentences uttered—already vindicated or—as in the case of truth, truthfulness, and rightness—could be vindicated because the sentences, propositions, expressed intentions, and utterances satisfy corresponding adequacy conditions.

Thus I distinguish (1) the *conditions* for the validity of a grammatical sentence, true proposition, truthful intentional expression, or normatively correct utterance suitable to its context, from (2) the *claims* with which speakers demand intersubjective recognition of the well-formedness of a sentence, truth of a proposition, truthfulness of an intentional expression, and rightness of a speech act, and from (3) the *vindication or redemption* of justified validity claims. Vindication means that the proponent, whether through appeal to intuitions and experiences or through argumentation and action consequences, grounds the claim's worthiness to be recognized [or acknowledged: *Anerkennungswürdigkeit*] and brings about a suprasubjective recognition of its validity. In accepting a validity claim raised by the speaker, the hearer acknowledges the validity of symbolic structures; that is, he acknowledges that a sentence is grammatical, a statement true, an intentional expression truthful, or an utterance correct. The validity of these symbolic structures is grounded in the fact that they satisfy certain adequacy conditions; but the meaning of the validity consists in worthiness to be recognized, that is, in the guarantee that intersubjective recognition can be brought about under suitable conditions.[4]

I have proposed the name *universal pragmatics* for the research program aimed at reconstructing the universal validity basis of speech.[5] I would like now to delimit the theme of this research program in a preliminary way. Thus before passing on (in part II) to the theory of speech acts, I shall prefix a few directorial remarks dealing with (1) a first delimitation of the object domain of the universal pragmatics called for; (2) an elucidation of the procedure of rational reconstruction, in contrast to empirical-analytic procedure in the narrower sense; (3) a few methodological difficulties resulting from the fact that linguistics claims the status of a reconstructive science; and finally (4) the question of whether the universal pragmatics proposed assumes the position of a transcendental reflective theory or that of a reconstructive science with empirical content. I shall restrict myself to directorial remarks because, while these questions are fundamental and deserve to be examined independently, they form only the context of the theme I shall treat and thus must remain in the background.

Preliminary Delimitation of the Object Domain

In several of his works, Apel has pointed to the abstractive fallacy that underlies the prevailing approach to the logic of science.[6] The logical analysis of language that originated with Carnap focuses primarily on syntactic and semantic properties of linguistic formations. Like structuralist linguistics, it delimits its object domain by first abstracting from the pragmatic properties of language, subsequently introducing the pragmatic dimension in such a way that the constitutive connection between the generative accomplishments of speaking and acting subjects, on the one hand, and the general structures of speech, on the other, cannot come into view. It is certainly legitimate to draw an abstractive distinction between language as structure and speaking as process. A language will then be understood as a system of rules for generating expressions, such that all well-formed expressions (e.g., sentences) may count as elements of this language. On the other hand, subjects capable of speaking can employ such expressions as participants in a process of communication; they can utter sentences as well as understand and respond to sentences expressed. This abstraction of *language* from the use of language in *speech* (*langue* versus *parole*), which is made in both the logical and the structuralist analysis of language, is meaningful. Nonetheless, this methodological step is not sufficient reason for the view that the pragmatic dimension of language from which one abstracts is beyond formal analysis. The fact of the successful, or at least promising, reconstruction of linguistic rule systems cannot serve as a justification for restricting formal analysis to this object domain. The separation of the two analytic levels, language and speech, should not be made in such a way that the pragmatic dimension of language is left to exclusively empirical analysis— that is, to empirical sciences such as psycholinguistics and sociolinguistics. I would defend the thesis that not only language but speech too—that is, the employment of sentences in utterances—is accessible to formal analysis. Like the elementary units of language (sentences), the elementary units of speech (utterances) can be analyzed in the methodological attitude of a reconstructive science.

Approaches to a general theory of communication have been developed from the *semiotics* of Charles Morris.[7] They integrate into their framework of fundamental concepts the model of linguistic behaviorism (the symbolically mediated behavioral reaction of the stimulated individual organism) and the model of information transmission (encoding and decoding signals between sender and receiver for a given channel and an at-least-partially-common store of signs). If the speaking process is thus conceptualized, the fundamental question of universal pragmatics concerning the general conditions of possible understanding cannot be suitably posed. For example, the intersubjectivity of meanings that are identical for at least two speakers does not even become a problem (1) if the identity of meanings is reduced to extensionally equivalent classes of behavioral properties, as is done in linguistic behaviorism;[8] or (2) if it is pre-established at the analytic level that there exists a common code and store of signs between sender and receiver, as is done in information theory.

In addition to empiricist approaches that issue, in one way or another, from the semiotics of Morris, there are interesting approaches to the formal analysis of general structures of speech and action. The following analyses can be understood as contributions along the way to a universal pragmatics. Bar Hillel pointed out quite early the necessity for a pragmatic extension of logical semantics.[9] Also of note are the proposals for a *deontic logic* (Hare, H. von Wright, N. Rescher)[10] and corresponding attempts at a formalization of such speech acts as commands and questions (Apostel);[11] approaches to a logic of nondeductive argumentation (Toulmin, Botha) belong here as well.[12] From the side of *linguistics*, the investigation of presuppositions (Kiefer, Petöfi),[13] conversational postulates (Grice, Lakoff),[14] speech acts (Ross, McCawley, Wunderlich),[15] and dialogues and texts (Fillmore, Posner)[16] lead to a consideration of the pragmatic dimension of language from a reconstructionist point of view. The difficulties in semantic theory (Katz, Lyons) point in the same direction.[17] From the side of *formal semantics,* the discussion—going back to Frege and Russell—of the structure of propositions, of referential terms and predicates (Strawson)[18] is particularly significant for a universal pragmatics. The same holds for analytic action theory (Danto, Hampshire, Schwayder)[19] and for the discussion that has arisen in connection with the logic of the explanation of intentional action (Winch, Taylor, von Wright).[20] The use theory of meaning introduced by Wittgenstein has universal-pragmatic aspects (Alston),[21] as does the attempt by Grice to trace meaning back to the intentions of the speaker (Bennett, Schiffer).[22] I shall draw primarily on the theory of speech acts initiated by Austin (Searle, Wunderlich),[23] which I take to be the most promising point of departure for a universal pragmatics.

These approaches developed from logic, linguistics, and the analytic philosophy of language have the common goal of clarifying processes of language use from the viewpoint of formal analysis. If one evaluates them with regard to the contribution they make to a universal pragmatics, their weaknesses also become apparent. In many cases, I see a danger that the analysis of conditions of possible understanding is cut short, either

a. Because these approaches do not generalize radically enough and do not push through the level of accidental contexts to general and unavoidable presuppositions—as is the case, for instance, with most of the linguistic investigations of semantic and pragmatic presuppositions; or

b. Because they restrict themselves to the instruments developed in logic and grammar, even when these are inadequate for capturing pragmatic relations—as, for example, in syntactic explanations of the performative character of speech acts; [24] or

c. Because they mislead one into a formalization of basic concepts that have not been satisfactorily analyzed—as can, in my view, be shown in the case of the logics of norms that trace norms of action back to commands; or finally,

d. Because they start from the model of the isolated, purposive-rational actor and thereby fail—as do, for example, Grice and Lewis[25] —to reconstruct in an appropriate way the specific moment of mutuality in the understanding of identical meanings or in the acknowledgment of intersubjective validity claims.

It is my impression that the theory of speech acts is largely free of these and similar weaknesses.

A Remark on the Procedure of Rational Reconstruction

I have been employing the expression *formal analysis* in opposition to *empirical-analytic procedures* (in the narrower sense) without providing a detailed explanation. This is at least misleading. I am not using formal analysis in a sense that refers, say, to the standard predicate logic or to any specific logic. The tolerant sense in which I understand formal analysis can best be characterized through the methodological attitude we adopt in the rational reconstruction of concepts, criteria, rules, and schemata. Thus we speak of the explication of meanings and concepts, of the analysis of presuppositions and rules. Of course, reconstructive procedures are also important for empirical-analytic research, for example, for explicating frameworks of basic concepts, for formalizing assumptions initially formulated in ordinary language, for clarifying deductive relations among particular hypotheses, for interpreting results of measurement, and so on. Nonetheless, reconstructive procedures are not characteristic of sciences that develop nomological hypotheses about domains of observable events; rather, these procedures are characteristic of *sciences that systematically reconstruct the intuitive knowledge of competent subjects.*

I would like to begin (clarifying the distinction between empirical-analytic and reconstructive sciences) with the distinction between sensory experience or *observation* and communicative experience or *understanding [Verstehen]*. Observation is directed to perceptible things and events (or states); understanding is directed to the meaning of utterances.[26] In experiencing, the observer is in principle alone, even if the categorial net in which experiences are organized with a claim to objectivity is already shared by several (or even all) individuals. In contrast, the interpreter who understands meaning is experiencing fundamentally as a participant in communication, on the basis of a symbolically established intersubjective relationship with other individuals, even if he is actually alone with a book, a document, or a work of art. I shall not here analyze the complex relationship between observation and understanding any further; but I would like to direct attention to one aspect—the difference in level between perceptible reality and the understandable meaning of a symbolic formation. Sensory experience is related to sectors of reality immediately, communicative experience only mediately, as illustrated in the diagram below.

Level 1 Observable Events ←—Observation (Observer)

Level 2   L – – – – –Observation Sentence ←—Understanding
                                                    (Interpreter)

Level 3   L – – – – – –Interpretation

This diagram represents three different relationships.

a. Epistemic relations between experiential acts and their objects. In this sense, the act of understanding relates to the symbolic expression (here of the observation sentence), as does the act of observation to the events observed.

b. Relations of representing an aspect of reality in a propositional sentence. In this sense, the interpretation represents the semantic content (here of the observation sentence), as the observation sentence in turn represents certain events.

c. Relations of expressing intentional acts. In this sense, the understanding (here of the observation sentence) is expressed in the propositional content of the interpretation, just as the observation is expressed in the propositional content of the observation sentence.

Apart from the fact that all three types of relation point to fundamental problems, there is an additional difficulty in specifying the precise differences between the epistemic relations of the observer and the interpreter to their respective objects and between the representational relation of the observation sentence to reality, on the one hand, and that of the interpretation sentence to *symbolically prestructured* reality, on the other. This specification would require a comparison between observation and interpretation, between description and explication. For the time being, the diagram merely illustrates the two levels of reality to which sensory and communicative experience relate. The difference in level between perceptible and symbolically prestructured reality is reflected in the gap between direct access through observation of reality and communicatively mediated access through understanding an utterance referring to events.

The two pairs of concepts—perceptible reality versus symbolically prestructured reality and observation versus understanding—can be correlated with the concepts of description versus explication. By using a sentence that reports an observation, I can describe the observed aspect of reality. By using a sentence that renders an interpretation of the meaning of a symbolic formation, I can explicate the meaning of such an utterance. Naturally only when the meaning of the symbolic formation is unclear does the explication need to be set off as an independent analytic step. In regard to sentences with which we describe events, there can be questions at different levels. If the phenomenon described needs explanation, we demand a causal description that makes clear how the phenomenon in question comes to pass. If, by contrast, the description itself is incomprehensible, we demand an explication that makes clear what the observer meant by his utterance and how the symbolic expression in need of elucidation comes about. In the first case, a satisfactory answer will have the form of an explanation we undertake with the aid of a causal hypothesis. In the second case, we speak of explication of meaning. (Of course, explications of meaning need not be limited to descriptive sentences; any meaningfully structured formation can be subjected to the operation of meaning explication.)

Descriptions and explications have different ranges; they can begin on the surface and push through to underlying structures. We are familiar with this fact in regard to the explanation of natural phenomena—theories can be more or less general. The same is true of meaning explications. Of course, the range of explication does not depend on the level of generality of theoretical knowledge about structures of an external reality accessible to observation but on knowledge of the deep structures of a reality accessible to understanding, the reality of symbolic formations produced according to rules. The explanation of natural phenomena pushes in a different direction from the explication of the meaning of expressions.

I want to distinguish two levels of explication of meaning. If the meaning of a written sentence, action, gesture, work of art, tool, theory, commodity, transmitted document, and so on, is unclear, the explication of meaning is directed first to the semantic content of the symbolic formation. In trying to understand its content, we take up the same position as the "author" adopted when he wrote the sentence, performed the gesture, used the tool, applied the theory, and so forth. Often too we must go beyond what was meant and intended by the author and take into consideration a context of which he was not conscious.[27] Typically, however, the *understanding of content* pursues connections that link the surface structures of the incomprehensible formation with the surface structures of other, familiar formations. Thus, linguistic expressions can be explicated through paraphrase in the same language or through translation into expressions of another language; in both cases, competent speakers draw on intuitively known meaning relations that obtain within the lexicon of one language or between those of two languages.

If he cannot attain his end in this way, the interpreter may find it necessary to alter his attitude. He then exchanges the attitude of understanding content—in which he looks, as it were, through symbolic formations to the world about which something is uttered—for an attitude in which he directs himself to the generative structures of the expressions themselves. The interpreter then attempts to explicate the meaning of a symbolic formation in terms of the rules according to which the author must have brought it forth. In normal paraphrase and translation, the interpreter draws on semantic meaning relations (for instance, between the different words of a language) in an ad hoc manner, in that he simply applies a knowledge shared with competent speakers of that language. In this sense, the role of interpreter can (under suitable conditions) be attributed to the author himself. The attitude changes, however, as soon as the interpreter tries not only to *apply* this intuitive knowledge but to *reconstruct* it. He then turns away from the surface structure of the symbolic formation; he no longer looks through it *intentione recta* to the world. He attempts instead to peer through the surface, as it were, and into the symbolic formation to discover the rules according to which the latter was produced (in our example, the rules according to which the lexicon of a language is constructed). The object of understanding is no longer the content of a symbolic expression or what specific authors meant by it in specific situations but the intuitive rule consciousness that a competent speaker has of his own language.

Borrowing from Ryle,[28] we can distinguish between *know-how* —the ability of a competent speaker who understands how to produce or perform something—and *know-that*—the explicit knowledge of how it is that he understands this. In our case, what the author means by an utterance and what an interpreter understands of its content, are a first-level know-that. To the extent that his utterance is correctly formed and thus comprehensible, the author produced it in accordance with certain rules or on the basis of certain structures. He understands the system of rules of his language and their context-specific application; he has a pre-theoretical knowledge of this rule system, which is at least sufficient to enable him to produce the utterance in question. This implicit rule consciousness is a know-how. The interpreter, in turn, who not only shares but wants to understand this implicit knowledge of the competent speaker, must transform this know-how into a second-level know-that. This is the task of reconstructive understanding, that is, of meaning explication in the sense of rational reconstruction of generative structures underlying the production of symbolic formations. Since the rule consciousness to be reconstructed is a categorial knowledge, the reconstruction first leads us to the operation of conceptual explication.

Carnap put forward four requirements, which the explication of a concept must fulfill in order to be adequate.

(1) The explicans should be *like* the explicandum, that is, from now on the explicans should be able to be used in place of the explicandum in all relevant cases. (2) There should be rules that fix the use of the explicans (in connection with other scientific concepts) in an exact manner. (3) The explicans should prove to be *fruitful* in regard to the formulation of general statements. (4) (Presupposing that requirements 1–3 can be met) the explicans should be as *simple* as possible.[29]

Wunderlich sums up his reflections on the status of concept explications as follows:

Explication always proceeds (conformable to Carnap's requirements 2–4) *with regard to theories;* either central concepts (like "meaning") are explicated in such a way that entire theories correspond to them as explicans, or different concepts are explicated interconnectedly. (2) We always explicate *with regard to clear cases,* so as to be able in connection with them to replace our intuitions with exact arguments. But the theory can then also provide answers to borderline cases; or we explicate separately what a clear borderline case is. (3) The language of explication is *at the same level* as the explicandum language (e.g., ordinary language or a standardized version derived from it). Thus it is not a question here of a descriptive language or a metalanguage relative to the language of the explicandum (the explicans does not describe the explicandum).[30]

In these reflections on the explication of concepts, one point strikes me as insufficiently worked out—*the evaluative accom-*

plishment of rule consciousness. Reconstructive proposals are directed to domains of pretheoretical *knowledge,* that is, not to any implicit opinion, but to a proven intuitive foreknowledge. The rule consciousness of competent speakers functions as a court of evaluation, for instance, with regard to the grammaticality of sentences. Whereas the understanding of content is directed to any utterance whatever, reconstructive understanding refers only to symbolic objects characterized as well formed by competent subjects themselves. Thus, for example, syntactic theory, propositional logic, the theory of science, and ethics start with syntactically well-formed sentences, correctly fashioned propositions, well-corroborated theories, and morally unobjectionable resolutions of norm conflicts, in order to reconstruct the rules according to which these formations can be produced. To the extent that universal-validity claims (the grammaticality of sentences, the consistency of propositions, the truth of hypotheses, the rightness of norms of action) underlie intuitive evaluations, as in our examples, reconstructions relate to pretheoretical knowledge of a general sort, to *universal capabilities,* and not only to particular competences of individual groups (e.g., the ability to utter sentences in a Low-German dialect or to solve problems in quantum physics) or to the ability of particular individuals (e.g., to write an exemplary *Entwicklungsroman* in the middle of the twentieth century). When the pretheoretical knowledge to be reconstructed expresses a universal capability, a general cognitive, linguistic, or interactive competence (or subcompetence), then what begins as an explication of meaning aims at the reconstruction of species competences. In scope and status, these reconstructions can be compared with general theories.[31]

It is the great merit of Chomsky to have developed this idea in the case of grammatical theory (for the first time in *Syntactic Structures,* 1957). Roughly speaking, it is the task of grammatical theory to reconstruct the rule consciousness common to all competent speakers in such a way that the proposals for reconstruction represent the system of rules that permits potential speakers to acquire the competence, in at least one language (*L*), to produce and to understand sentences that count as grammatical in *L,* as well as to distinguish sentences well-formed in *L* from ungrammatical sentences.[32]

Reconstructive versus Empiricist Linguistics

I hope I have characterized the reconstructive procedure of sciences that transform a practically mastered pretheoretical knowledge (know-how) of competent subjects into an objective and explicit knowledge (know-that) to an extent sufficient to make clear in what sense I am using the expression *formal analysis*. Before mentioning some methodological difficulties with reconstructive linguistics, I would like to contrast, in broad strokes, two versions of the science of language, one empirical-analytic and the other reconstructive. (Wunderlich speaks of empirical-descriptive and empirical-explicative science of language.[33])

Data. To the extent that the experiential basis is supposed to be secured through observation alone, the data of linguistics consist of measured variables of linguistic behavior. By contrast, to the extent that reconstructive understanding is permitted, the data are provided by the rule consciousness of competent speakers, maeutically ascertained (i.e., through suitable questioning with the aid of systematically ordered examples). Thus the data are distinguished, if you will, by their ontological level: actual linguistic behavior is part of perceptible reality, and rule-consciousness points to the production of symbolic formations in which something is uttered about reality.[34] Furthermore, observations always mean a knowledge of something particular, whereas rule consciousness contains categorical knowledge. Finally, observational data are selected only from the analytic viewpoints of the linguist, whereas, in the other case, competent speakers themselves evaluate and preselect possible data from the point of view of their grammatical well-formedness.

Theory and Object Domain. As long as natural languages count as the object of linguistic description and not as the form of representation of a reconstructible pretheoretical knowledge, linguistic theory relates to its object domain as an empirical theory that explains linguistic descriptions of linguistic reality with the aid of nomological hypotheses. If, on the contrary, linguistic theory is supposed to serve to reconstruct pretheoretical knowledge, theory relates to its object domain as an explication of meaning to its explicandum. Whereas in the empiricist version the relation of theory to the language to be explained is basically indistinguishable from that between theory and reality in other nomological sciences, in the explicative version the linguistic character of the object necessitates a relation that can hold only between different linguistic expressions: the relation between explication and explicandum, where the language of explication (that is, the construct language of linguistic science, which is a standardized version of ordinary language) belongs in principle to the same level as the natural language to be explicated. (Neither in the descriptive nor in the explicative case of theory formation can the relation of linguistic theory to its object domain be conceived as that of metalanguage to object language.)[35]

Theory and Everyday Knowledge. There is yet another peculiarity arising from these differently oriented conceptualizations. An empirical-analytic theory in the narrow sense can (and as a rule will) refute the everyday knowledge of an object domain that we possess prior to science and replace it with a correct theoretical knowledge regarded provisionally as true. A proposal for reconstruction, by contrast, can represent pretheoretical knowledge more or less explicitly and adequately, but it can never falsify it. At most, the report of a speaker's intuition can prove to be false, but not the intuition itself.[36] The latter belongs to the data, and data can be explained but not criticized. At most, data can be criticized as being unsuitable, that is, either erroneously gathered or wrongly selected for a specific theoretical purpose.

To a certain extent, reconstructions make an essentialist claim. Of course, one can say that theoretical descriptions correspond (if true) to certain structures of reality in the same sense as reconstructions bear a likeness (if correct) to the deep structures explicated. On the other hand, the asserted correspondence between a descriptive theory and an object allows of many epistemological interpretations other than the realistic (e.g., instrumentalist or conventionalist). Rational reconstructions, on the contrary, can reproduce the pretheoretical knowledge that they explicate only in an essentialist sense; if they are true, they have to correspond precisely to the rules that are operatively effective in the object domain—that is, to the rules that actually determine the production of surface structures.[37] Thus Chomsky's correlation assumption, according to which linguistic grammar is represented on the part of the speaker by a corresponding mental grammar is, at least in the first instance, consistent.

Methodological Difficulties. To be sure, serious methodological difficulties have arisen in connection with the Chomskian program for a general science of language as the rational reconstruction of linguistic competence. I would like to consider, from a method- ological perspective, two of the problem complexes that have de- veloped. One concerns the status and reliability of the intuitive knowledge of competent speakers; the other, the aforementioned relation between linguistic and mental grammar.

There have been above all two objections against choosing speakers' intuitions as the starting point of reconstructive theory formation.[38] First, the question has been raised whether a re- constructive linguistics can ever arrive at a theory of linguistic competence, whether on the chosen data basis it is not, rather, limited to developing, at best, a theory of the intuitive under- standing that competent speakers have of their own language? Since the metalinguistic use of one's own ordinary language, to which a science that appeals to speakers' judgments must have recourse, is something other than the direct use of language (and is probably subject to different laws), a grammatical theory of the Chomskian type can reconstruct, at best, that special part of linguistic competence that rules the metalinguistic use; it cannot reconstruct the competence that directly underlies speaking and understanding a language.

The empirical question is whether a complete theory of linguistic intu- itions is identical with a complete theory of human linguistic compe- tence. . . . Chomsky has no doubt as to this identity. . . . The theory of one kind of linguistic behavior, namely metalinguistic judgment on such things as grammaticality and paraphrase, would then as a whole be built into theories on other forms of linguistic behavior such as speaking and understanding. . . . If we wish to think in terms of pri- mary and derived forms of verbal behavior, the speaking and the understanding of language fall precisely into the category of primary forms, while metalinguistic judgments will be considered highly de- rived, artificial forms of linguistic behavior, which moreover are acquired late in development. . . . The empirical problem in the psychology of language is in turn divided in two, the investigation of psychological factors in primary language usage, and the psychological investigation of linguistic intuitions.[39]

I think this objection is based on a confusion of the two research paradigms elucidated above, the empirical-analytic and the reconstructive, to which I shall address the three following remarks.

1. Reconstruction relates to a pretheoretical knowledge of competent speakers that is expressed in the production of sentences in a natural language, on the one hand, and in the appraisal of the grammaticality of linguistic expressions, on the other. The object of reconstruction is the process of production of sentences held by competent speakers to belong to the set of grammatical sentences. The metalinguistic utterances in which competent speakers evaluate the sentences put before them are not the object of reconstruction but part of the data gathering.

2. Because of the reflexive character of natural languages, speaking about what has been spoken, direct or indirect mention of speech components, belongs to the normal linguistic process of reaching understanding. The expression *metalinguistic judgments* in a natural language about sentences of the same language suggests a difference of level that does not exist. It is one of the interesting features of natural languages that they can be used as their own language of explication. (I shall come back to this point below.)

3. However, it seems to me that the misunderstanding lies, above all, in Levelt's considering the recourse to speakers' intuitions in abstraction from the underlying research paradigm. Only if one presupposes an empirical-analytic (in the narrow sense) approach to the reality of a natural language and the utterances in it, can one view speaking and understanding language, on the one hand, and judgments in and about a language, on the other, as two different object domains. If one chooses a reconstructive approach, then one *thereby* chooses a conceptualization of the object domain according to which the linguistic know-how of a competent speaker is at the root of the sentences he produces with the help of (and only with the help of) this know-how. While this research paradigm may prove to be unfruitful, this cannot be shown at the level of a critique that already presup-

poses a competing paradigm; it has to be shown in terms of the
success or failure of the theories and explanations the competing
paradigms make possible.

The second objection is directed toward the unreliability of
intuitively founded speakers' intuitions, for which there exists
impressive empirical evidence.[40] Here again, it seems to me that
an empiricist interpretation of speakers' judgments stimulates
false expectations and suggests the wrong remedy. The expression
intuitive knowledge should not be understood as meaning that a
speaker's pretheoretical knowledge about the grammaticality of a
sentence (the rigor of a derivation, the cogency of a theory, and
so forth) is the kind of directly ascertainable intuition that is
incapable of being discursively justified. On the contrary, the
implicit knowledge has to be brought to consciousness through
the choice of suitable examples and counterexamples, through
contrast and similarity relations, through translation, paraphrase,
and so on—that is, through a well-thought-out maeutic method
of interrogation. Ascertaining the so-called intuitions of a speaker
is already the beginning of their explication. For this reason, the
procedure practiced by Chomsky and many others seems to me to
be meaningful and adequate. One starts with clear cases, in which
the reactions of the subjects converge, in order to develop structural descriptions on this basis and then, in the light of the hypotheses gained, to present less clear cases in such a way that the
process of interrogation can lead to an adequate clarification of
these cases as well. I do not see anything wrong in this circular procedure; every research process moves in such a circle
between theory formation and precise specification of the object
domain.[41]

The second methodological question is more difficult. It is one
that has been treated as an empirical question in the psycholinguistics of the past decade, and as such has inspired a great
amount of research: is there a direct correspondence between the
linguistic theory of grammar and the mental grammar that is, so
to speak, "in the mind" of the speaker? [42] According to the correlation hypothesis, linguistic reconstructions are not simply lucid
and economical representations of linguistic data; instead, there
is a psychological complexity of the actual production process that

corresponds, supposedly, to the transformational complexity that can be read off the structural description of linguistic expressions. I cannot go into the individual research projects and the different interpretations here. Apparently in psycholinguistics there is a growing tendency to disavow the original correlation hypothesis; the mental grammar that underlies the psychologically identifiable production of language and the corresponding processes of understanding cannot, in the opinion of Bever, Watt, and others, be explained in the framework of a competence theory, that is, of a reconstructively oriented linguistics. I am not very certain how to judge this controversy; but I would like to suggest two points of view that have not, so far as I can see, been taken sufficiently into account in the discussion.

1. How strong do the essentialist assertions of a reconstructive linguistics regarding the psychic reality of reconstructed systems of rules have to be? Chomsky's maturationist assumption—that grammatical theory represents exactly the innate dispositions that enable the child to develop the hypotheses that direct language acquisition and to process the linguistic data in the environment— seems to me too strong.[43] Within the reconstructivist conceptual strategy, the more plausible assumption that grammatical theory represents the linguistic competence of the adult speaker is sufficient. This competence in turn is the result of a learning process that may—like cognitive development in the sense of Piaget's cognitivist approach—follow a rationally reconstructible pattern.[44] As Bever suggests, even this thesis can be weakened to allow for the limitations placed on the acquisition and application of grammatical-rule knowledge by nonlinguistic perceptual mechanisms or nonlinguistic epistemic systems in general, without surrendering the categorial framework of a competence theory.

2. It is not clear to me to what extent the psycholinguistic critique of the admittedly essentialist implications of Chomsky's competence theory originates in a confusion of research paradigms. This could be adequately discussed only if there were clarity about the way in which competence theories can be tested and falsified. I have the impression that psycholinguistic investigations proceed empirical-analytically and neglect *a limine* the distinction between competence and performance.[45]

Universal Pragmatics versus Transcendental Hermeneutics

Having presented the idea of a reconstructive science and briefly elucidated it through a consideration of reconstructive linguistics (and two of its methodological difficulties), I would like to pose one further question: what is the relation of universal-pragmatic reconstruction of general and unavoidable presuppositions of possible processes of understanding to the type of investigation that has, since Kant, been called transcendental analysis? Kant terms *transcendental* an investigation that identifies and analyzes the a priori conditions of possibility of experience. The underlying idea is clear: in addition to the empirical knowledge that relates to objects of experience, there is, supposedly, a transcendental knowledge of concepts of objects in general that precedes experience. The method by which these a priori concepts of objects in general can be shown to be valid conditions of possible experience is less clear. There is already disagreement concerning the meaning of the thesis: "the a priori conditions of a possible experience in general are at the same time conditions of the possibility of objects of experience."[46]

The analytic reception of the Kantian program (Strawson's work is a familiar example)[47] leads to a minimalist interpretation of the transcendental. Every coherent experience is organized in a categorial network; to the extent that we discover the same implicit conceptual structure in any coherent experience whatsoever, we may call this basic conceptual system of possible experience *transcendental*. This conception renounces the claim that Kant wanted to vindicate with his transcendental deduction; it gives up all claim to a proof of the objective validity of our concepts of objects of possible experience in general.[48] The strong apriorism of Kantian philosophy gives way to a weaker version. From now on, transcendental investigation must rely on the competence of knowing subjects who judge which experiences may be called coherent experiences in order to analyze this material for general and necessary categorial presuppositions. Every reconstruction of a basic conceptual system of possible experience has to be regarded as a hypothetical proposal that can be tested against new experiences. As long as the assertion of its necessity and universality has not been refuted, we term *transcendental* the conceptual structure recurring in all coherent experiences. In this weaker version, the claim that that structure can be demonstrated a priori is dropped.

From this modification follow consequences that are scarcely compatible with the original program. We can no longer exclude the possibility that our concepts of objects of possible experience can be successfully applied only under contingent boundary conditions that, let us say, have heretofore been regularly fulfilled by natural constants.[49] We can no longer exclude the possibility that the basic conceptual structure of possible experience has developed phylogenetically and arises anew in every normal ontogenesis, in a process that can be analyzed empirically.[50] We cannot even exclude the possibility that an a priori of experience that is relativized in this sense is valid only for specific, anthropologically deep-seated behavioral systems, each of which makes possible a specific statategy for objectivating reality. The transcendentally oriented pragmatism inaugurated by C. S. Peirce attempts to show that there is such a structural connection between experience and instrumental action;[51] the hermeneutics stemming from Dilthey attempts—over against this a priori of experience— to do justice to an additional a priori of understanding or communicative action.[52]

From the perspective of a transformed transcendental philosophy (in Apel's sense), two further renunciations called for by the analytic reception of Kant seem precipitate: the renunciation of the concept of the constitution of experience and the renunciation of an explicit treatment of problems of validity. In my opinion, the reservation regarding a strong apriorism in no way demands limiting oneself to a logical-semantic analysis of the conditions of possible experiences. If we surrender the concept of the transcendental subject—the subject that accomplishes the synthesis and that, together with its knowledge-enabling structures, is removed from all experience—this does not mean that we have to renounce universal-pragmatic analysis of the application of our concepts of objects of possible experience, that is, investigation of the constitution of experience.[53] It is just as little a consequence of giving up the project of a transcendental deduction that one must hand over problems of validity to other domains of investigation, say to the theory of science or of truth. Of course, the relation between the objectivity of possible experience and the truth of propositions looks different than it does under Kantian premises. In place of a priori demonstration, we have transcendental investigation of the conditions for argumentatively redeeming validity claims that are at least implicitly related to discursive vindication.[54]

In my view, it is not merely a terminological question whether we call such investigations of general and unavoidable presuppositions of communication (in this case, presuppositions of argumentative speech) *transcendental*. If we want to subject processes of reaching understanding (speech) to a reconstructive analysis oriented to general and unavoidable presuppositions in the same way as has been done for cognitive processes,[55] then the model of transcendental philosophy undeniably suggests itself, all the more so as the theory of language and action has not (despite Humboldt) found its Kant. Naturally, recourse to this model is understandable only if one has in view one of the weaker versions of transcendental philosophy mentioned above. In this sense, Apel speaks of "transcendental hermeneutics" or "transcendental pragmatics" in order to characterize his approach programmatically. I would like to mention two reasons for hesitating to adopt this usage.

a. Something like a transcendental investigation of processes of understanding seems plausible to me as long as we view these under the aspect of processes of experience. It is in this sense that I speak of communicative experience; in understanding the utterance of another speaker as a participant in a communication process, the hearer (like the observer who perceives a segment of reality) has an experience. From this comparative perspective, concrete utterances would correspond to empirical objects, and utterances in general to objects in general (in the sense of objects of possible experience). Just as we analyze our a priori concepts of objects in general—that is, the conceptual structure of any coherent perception—we could analyze our a priori concepts of utterances in general—that is, the basic concepts of situations of possible understanding, the conceptual structure that enables us to employ sentences in correct utterances. Concepts such as meaning and intentionality, the ability to speak and act (agency), interpersonal relation, and the like, would belong to this conceptual framework.

The expression "situation of possible understanding" that would correspond to the expression "object of possible experience" from this point of view, already shows, however, that acquiring the experiences we have in processes of communication is secondary to the goal of reaching understanding that these processes serve. The general structures of speech must first be investigated from the perspective of understanding and not from that of experience. As soon as we admit this, however, the parallels with transcendental philosophy (however conceived) recede into the background. The idea underlying transcendental philosophy is—to oversimplify—that we constitute experiences in objectivating reality from invariant points of view; this objectivation shows itself in the objects in general that are necessarily presupposed in every coherent experience; these objects in turn can be analyzed as a system of basic concepts. However, I do not find any correspondent to this idea under which the analysis of general presuppositions of communication might be carried out. Experiences are, if we follow the basic Kantian idea, constituted; utterances are at most generated. A transcendental investigation transposed to processes of understanding would thus have to be oriented around another model—not the epistemological model of the constitution of experience but perhaps the model of deep and surface structure.

b. Moreover, adopting the expression *transcendental* could conceal the break with apriorism that has been made in the meantime. Kant had to separate empirical from transcendental analysis sharply. If we now understand transcendental investigation in the sense of a reconstruction of general and unavoidable presuppositions of experiences that can lay claim to objectivity, then there certainly remains a difference between reconstructive and empirical-analytic analysis. But the distinction between drawing on a priori knowledge and drawing on a posteriori knowledge becomes blurred. On the one hand, the rule consciousness of competent speakers is for them an a priori knowledge; on the other

hand, the reconstruction of this knowledge calls for inquiries undertaken with empirical speakers—the linguist procures for himself a knowledge a posteriori. The implicit knowledge of competent speakers is so different from the explicit form of linguistic
description that the individual linguist cannot rely on reflection on
his own speech intuitions. The procedures employed in constructing and testing hypotheses, in appraising competing reconstructive proposals, in gathering and selecting data, are in
many ways like the procedures used in the nomological sciences.
Methodological differences that can be traced back to differences
in the structure of data (observable events versus understandable
signs) and to differences between the structures of laws and rules,
do not suffice to banish linguistics, for example, from the sphere
of empirical science.

This is particularly true of ontogenetic theories that, like
Piaget's cognitive developmental psychology, connect the structural description of competences (and of reconstructed patterns
of development of these competences) with assumptions concerning causal mechanisms.[56] The paradigms introduced by
Chomsky and Piaget have led to a type of research determined
by a peculiar connection between formal and empirical analysis
rather than by their classical separation. The expression *transcendental,* with which we associate a contrast to empirical science, is thus unsuited to characterizing, without misunderstanding,
a line of research such as universal pragmatics. Behind the terminological question, there stands the systematic question concerning the as-yet insufficiently clarified status of nonnomological
empirical sciences of the reconstructive type. I shall have to leave
this question aside here. In any case, the attempt to play down
the interesting methodological differences that arise here, and
to interpret them away in the sense of the unified science program,
seems to have little prospect of success.[57]

II

The discussion of the theory of speech acts has given rise to ideas
on which the fundamental assumptions of universal pragmatics
can be based.[58] The universal-pragmatic point of view from which

I shall select and discuss these ideas leads, however, to an interpretation that diverges in several important respects from the
understanding of Austin and Searle.

Three Aspects of Universal Pragmatics

The basic universal-pragmatic intention of speech-act theory is
expressed in the fact that it thematizes the elementary units of
speech (utterances) in an attitude similar to that in which linguistics does the units of language (sentences). The goal of reconstructive language analysis is an explicit description of the
rules that a competent speaker must master in order to form
grammatical sentences and to utter them in an acceptable way.
The theory of speech acts shares this task with linguistics.
Whereas the latter starts from the assumption that every adult
speaker possesses an implicit, reconstructible knowledge, in which
is expressed his linguistic rule competence (to produce sentences),
speech-act theory postulates a corresponding communicative rule
competence, namely the competence to employ sentences in speech
acts. It is further assumed that communicative competence has
just as universal a core as linguistic competence. A general theory
of speech actions would thus describe exactly that fundamental
system of rules that adult subjects master to the extent that they
can fulfill *the conditions for a happy employment of sentences
in utterances,* no matter to which particular language the sentences may belong and in which accidental contexts the utterances
may be embedded.

The proposal to investigate language use in competence-theo-
retic terms calls for a revision of the concepts of competence and
performance. Chomsky understands these concepts in such a way
that it makes sense to require that phonetic, syntactic, and semantic properties of sentences be investigated linguistically within the
framework of a reconstruction of linguistic competence and that
pragmatic properties of utterances be left to a theory of linguistic
performance.[59] This conceptualization gives rise to the question
of whether communicative competence is not a hybrid concept. I
have, to begin with, based the demarcation of linguistics from
universal pragmatics on the current distinction between sentences

and utterances. The production of sentences according to the rules of grammar is something other than the use of sentences in accordance with pragmatic rules that shape the infrastructure of speech situations in general. But this raises the following questions. (1) Could not the universal structures of speech—what is common to all utterances independently of their particular contexts—be adequately determined through universal sentential structures? In this case, with his linguistically reconstructible linguistic competence, the speaker would also be equipped for mastering situations of possible understanding, for the general task of uttering sentences; and the postulate of a general communicative competence different from the linguistic could not be justified. Beyond this there is the question, (2) whether the semantic properties of sentences (or words), in the sense of the use theory of meaning, can in any case be explicated only with reference to situations of possible typical employment. Then the distinction between sentences and utterances would be irrelevant, at least to semantic theory (as long, at any rate, as sufficiently typical contexts of utterance were taken into consideration). As soon as the distinction between the linguistic analysis of sentences and the pragmatic analysis of utterances becomes hazy, the object domain of universal pragmatics is in danger of fading away.

[In reference to question 1,] I would agree, with certain qualifications,[60] with the statement that a speaker, in transposing a well-formed sentence into an act oriented to reaching understanding, merely actualizes what is inherent in the sentence structures. But this is not to deny the difference between the production of a grammatical sentence and the use of that sentence in a situation of possible understanding, or the difference between the universal presuppositions that a competent speaker has to fulfill in each case. In order to utter a sentence, the speaker must fulfill general presuppositions of communication. Even if he fulfills these presuppositions in conformity to the structures that are already given with the sentence employed, he may very well form the sentence itself without also fulfilling the presuppositions specific to the telos of communication. This can be made clear with regard to the relations to reality in which every sentence is first embedded through the act of utterance. In being uttered, a sentence is placed in relation to (1) the external reality of what is supposed to be an existing state of affairs, (2) the internal reality of what a speaker would like to express before a public as his intentions, and, finally, (3) the normative reality of what is intersubjectively recognized as a legitimate interpersonal relationship. It is thereby placed under validity claims that it need not and cannot fulfill as a nonsituated sentence, as a purely grammatical formation. A chain of symbols "counts" as a sentence of a natural language, L, if it is well-formed according to the system of grammatical rules, GL. The grammaticality of a sentence means (from a pragmatic perspective) that the sentence, when uttered by a speaker, is *comprehensible* to all hearers who have mastered GL. Comprehensibility is the only one of these universal claims that can be fulfilled immanently to language. The validity of a propositional content depends, by contrast, on whether the proposition stated represents a fact (or whether the existential presuppositions of a mentioned propositional content hold); the validity of an intention expressed depends on whether it corresponds to what is actually intended by the speaker; and the validity of the utterance performed depends on whether this action conforms to a recognized normative background. Whereas a grammatical sentence fulfills the claim to comprehensibility, a successful utterance must satisfy three additional validity claims: it must count as true for the participants insofar as it represents something in the world; it must count as truthful insofar as it expresses something intended by the speaker; and it must count as right insofar as it conforms to socially recognized expectations.

Naturally we can identify characteristics of the surface structures of sentences that have a special significance for the three general pragmatic functions of the utterance: to represent something, to express an intention, to establish a legitimate interpersonal relation. Propositional sentences can be used to represent an existing state of affairs (or to mention them indirectly in non-constative speech acts); intentional verbs, modal forms, and so on can be used to express the speaker's intentions; performative phrases, illocutionary indicators, and the like can be used to establish interpersonal relations between speaker and hearer. Thus the general structures of speech are also reflected at the level of sentence structure. But insofar as we consider a sentence as a grammatical formation, that is, as independent of speech situations in which it can be uttered, these general pragmatic functions are not yet "occupied." To bring forth a grammatical sentence—as an example, say, for linguists—a competent speaker need satisfy only the claim to comprehensibility. He has to have mastered the corresponding system of grammatical rules; this we call his linguistic ability, and it can be analyzed linguistically. It is otherwise with his ability to communicate; this is susceptible only to pragmatic analysis. By "communicative competence" I understand the ability of a speaker oriented to mutual understanding to embed a well-formed sentence in relations to reality, that is:

1. To choose the propositional sentence in such a way that either the truth conditions of the proposition stated or the existential presuppositions of the propositional content mentioned are supposedly fulfilled (so that the hearer can share the knowledge of the speaker);

2. To express his intentions in such a way that the linguistic expression represents what is intended (so that the hearer can trust the speaker);

3. To perform the speech act in such a way that it conforms to recognized norms or to accepted self-images (so that the hearer can be in accord with the speaker in shared value orientations).

To the extent that these decisions do not depend on particular epistemic presuppositions and changing contexts but cause sentences in general to be engaged in the universal pragmatic functions of representation, expression, and legitimate interpersonal relation, what is expressed in them is precisely the communicative competence for which I am proposing a universal-pragmatic investigation.

The part of universal pragmatics that is furthest developed is that related to the representational function of utterances, for example, to the use of elementary propositional sentences. This classic domain of formal semantics has been pursued from Frege to Dummet.[61] That this is a matter of universal-pragmatic investigation can be seen in the fact that the truth value of propositions is systematically taken into account. The theory of predication does not investigate sentences in general (as does linguistics) but sentences in their function of representing facts. The analysis is directed above all to the logic of using predicates and those expressions that enable us to refer to objects. Naturally this part of universal pragmatics is not the most important for a theory of communication. The analysis of intentionality, the discussion of avowals, and the debate on private speech, insofar as they clear the way to a universal pragmatics of the expressive function of utterances, are only beginnings.[62] Finally, speech-act theory provides a point of departure for the part of universal pragmatics related to the interpersonal function of utterances.

[In reference to question 2,] one might see a further difficulty with my proposal for conceptualizing universal pragmatics in the fact that formal semantics does not fit well into the distinction between a linguistic analysis concerned with sentences and a pragmatic analysis concerned with utterances. There is a broad spectrum of different approaches to semantic theory. Linguistically oriented theories of meaning[63] try to grasp systematically the semantic content of linguistic expressions. In the framework of transformational grammar, explanations of the surface structures of sentences either start with semantic deep structures or rely on semantic projections into syntactic structures. This approach leads to an elementaristically constructed combinatory system of general semantic markers. Lexical semantics proceeds in a similar manner; it clarifies the meaning structures of a given lexicon by way of a formal analysis of meaning relations. The weakness of these linguistic approaches lies in the fact that they bring in the pragmatic dimension of the use of sentences only in an ad hoc way. The use theory of meaning developed from the work of Wittgenstein has shown, however, that the meaning of linguistic expressions can be identified only with reference to situations of possible employment.

For their part, *pragmatic theories of semantics*[64] are faced with the difficulty of delimiting a linguistic expression's typical situations of employment from contexts that happen by chance to have additional meaning-generating power but do not affect the semantic core of the linguistic expression. *Reference semantics,*[65] whether framed as a theory of extensional or of intensional denotation, determines the meaning of an expression by the class of objects to which it can be applied in true sentences. Under these premises one can explicate the meaning of expressions that appear in propositional sentences. I do not see, however, why semantic theory should monopolistically single out the representational function of language and ignore the specific meanings that language develops in its expressive and interpersonal functions.

These preliminary reflections are intended merely to support the conjecture that semantic theory cannot be successfully developed as a unified theory. But if it is heterogeneously composed, no objection to the methodological separation of the analysis of sentence structures from that of utterance structures can be inferred from the difficulties of demarcating semantics from pragmatics (difficulties that are equally present in demarcating semantics from syntax). The analysis of general structures of speech can indeed begin with general sentence structures. However, it is directed to formal properties of sentences only from the perspective of the possibility of *using sentences* as elements of speech, that is, for representational, expressive, and interpersonal functions. Universal pragmatics too can be understood as semantic analysis. But it is distinguished from other theories of meaning in that the meanings of linguistic expressions are relevant only insofar as they contribute to speech acts that satisfy the validity claims of truth, truthfulness, and normative rightness. On the other hand, universal pragmatics is distinguished from empirical pragmatics, e.g., sociolinguistics, in that the meaning of linguistic expressions comes under consideration only insofar as it is determined by formal properties of speech situations in general, and not by particular situations of use.

I would like now to sum up the different levels of analysis and corresponding object domains of semiotics.

a. *Sentences versus Utterances.* If we start with concrete speech actions embedded in specific contexts and then disregard all aspects that these utterances owe to their pragmatic functions, we are left with linguistic expressions. Whereas the elementary unit of speech is the speech act, the elementary unit of language is the sentence. The demarcation is obtained by attending to conditions of validity—a grammatically well-formed sentence satisfies the claim to comprehensibility;

a communicatively successful speech action requires, beyond the comprehensibility of the linguistic expression, that the participants in communication be prepared to reach an understanding, that they raise claims to truth, truthfulness, and rightness and reciprocally impute their satisfaction. Sentences are the object of linguistic analysis (b, c), speech acts of pragmatic analysis (d, e).

b. *Individual Languages versus Language in General.* The task of linguistics consists firstly in developing a grammar for each individual language so that a structural description can be correlated with any sentence of the language. On the other hand, general grammatical theory is occupied with reconstructing the rule system that underlies the ability of a subject to generate well-formed sentences in any language whatever. Grammatical theory claims to reconstruct the universal linguistic ability of adult speakers. (In a strong version, this linguistic competence means the ability to develop hypotheses that guide language acquisition on the basis of an innate disposition; in a weaker version, linguistic competence represents the result of learning processes interpreted constructivistically in Piaget's sense.)

c. *Aspects of Linguistic Analysis.* Every linguistic utterance can be examined from at least three analytic viewpoints. Phonetics investigates linguistic expressions as inscriptions in an underlying medium (i.e., as formations of sound). Syntactic theory investigates linguistic expressions with regard to the formal connections of the smallest meaningful units. Semantic theory investigates the meaning content of linguistic expressions. Evidently only phonetic and syntactic theory are self-sufficient linguistic theories, whereas semantic theory cannot be completely carried through in the attitude of the linguist, that is, in disregard of pragmatic aspects.

d. *Particular versus Universal Aspects of Speech Acts.* The task of empirical pragmatics consists, to begin with, in describing speech acts typical of a certain milieu, which can in turn be analyzed from sociological, ethnological, and psychological points of view. General pragmatic theory, on the other hand, is occupied with reconstructing the rule system that underlies the ability of a subject to utter sentences in any relevant situation. Universal pragmatics thereby raises the claim to reconstruct the ability of adult speakers to embed sentences in relations to reality in such a way that they can take on the general pragmatic functions of representation, expression, and establishing legitimate interpersonal relations. This communicative competence is indicated by those accomplishments that hermeneutics stylizes to an art, namely paraphrasing utterances by means of context-similar utterances of the same language or translating them into context-comparable utterances in a foreign language.

e. *Universal-Pragmatic Aspects.* The three general pragmatic functions—with the help of a sentence, to represent something in the world, to express the speaker's intentions, and to establish legitimate interpersonal relations—are the basis of all the particular functions that an utterance can assume in specific contexts. The fulfillment of those general functions is measured against the validity conditions for truth, truthfulness, and rightness. Thus every speech action can be considered from the corresponding analytic viewpoints. Formal semantics examines the structure of elementary propositions and the acts of reference and predication. A still scarcely developed theory of intentionality examines intentional expressions insofar as they function in first-person sentences. Finally, the theory of speech acts examines illocutionary force from the viewpoint of the establishment of legitimate interpersonal relations. These semiotic distinctions are summarized in the following table.

Theoretical Level	Object Domain
Linguistics	Sentences
Grammar	Sentences of an individual language
Grammatical theory	Rules for generating sentences in any language whatever
Aspects of linguistic analysis	
Phonetic theory	Inscriptions (language sounds)
Syntactic theory	Syntactical rules
Semantic theory	Lexical units
Pragmatics	Speech actions
Empirical pragmatics	Context-bound speech actions
Universal pragmatics	Rules for using sentences in utterances
Aspects of universal-pragmatic analysis	
Theory of elementary propositions	Acts of reference and predication
Theory of first-person sentences	Linguistic expression of intentions
Theory of illocutionary acts	Establishment of interpersonal relations

For a theory of communicative action, the third aspect of utterances, namely the establishment of interpersonal relations, is central. I shall therefore take the theory of speech acts as my point of departure.

The Standard Form of the Speech Act—
Searle's Principle of Expressibility

The principal task of speech-act theory is to clarify the performative status of linguistic utterances. Austin analyzed the sense in which I can utter sentences in speech acts as the *illocutionary force* of speech actions. In uttering a promise, an assertion, or a warning, together with the corresponding sentences I execute an action—I try to *make* a promise, to *put forward* an assertion, to *issue* a warning—I do things by saying something. Although there are other modes of employing language—Austin mentions, among others, writing poems and telling jokes—the illocutionary use seems to be the foundation on which even these other kinds of employment rest. To be understood in a given situation, every utterance must, at least implicitly, establish and bring to expression a certain relation between the speaker and his counterpart. We can also say that the illuocutionary force of a speech action consists in fixing the communicative function of the content uttered.

The current distinction between the content and the relational aspects of an utterance has, to begin, a trivial meaning.[66] It says that in being uttered the sentence is embedded in specific interpersonal relations. In a certain way, every explicitly performative utterance both establishes and represents an interpersonal relation. This circumstance is trivial so long as under the relational aspect we merely contrast the utterance character of speech with its semantic content. If nothing more were meant by the illocutionary force of a speech act, the concept "illocutionary" could serve at best to elucidate the fact that linguistic utterances have the character of actions, that is, are speech actions. The point of the concept cannot lie therein; I find it rather in the peculiarly generative power of speech acts.

It is to this generative power that I trace the fact that a speech act can succeed (or fail). We can say that a speech act succeeds if a relation between the speaker and the hearer comes to pass—indeed the relation intended by the speaker—and if the hearer can *understand and accept* the content uttered by the speaker in the sense indicated (e.g., as a promise, assertion, suggestion, and so forth). Thus the generative power consists in the fact that the speaker, in performing a speech act, can influence the hearer in such a way that the latter can take up an interpersonal relation with him.[67] It can, of course, be said of every interaction, and not only of speech actions, that it establishes an interpersonal relation. Whether or not they have an explicitly linguistic form, communicative actions are related to a context of action norms and values. Without the normative background of routines, roles, forms of life—in short, conventions—the individual action would remain indeterminate. All communicative actions satisfy or violate normative expectations or conventions. Satisfying a convention in acting means that a subject capable of speaking and acting takes up an interpersonal relation with at least one other such subject. Thus the establishment of an interpersonal relation is a criterion that is not selective enough for our purposes. I emphasized at the start that I am restricting my analysis to paradigmatic cases of linguistically explicit action that is oriented to reaching understanding. This restriction must now be drawn somewhat more precisely. In doing so, we can begin with the standard examples from which speech-act theory was developed. The following are typical speech-act forms:[68]

"I . . .   you that . . .   ."
    [verb]   [sentence]
e.g., "I (hereby) promise you that I will come tomorrow."

"You are . . .   ."
            [verb] [p. part.]   [sentence]
e.g., "You are requested to stop smoking."

"I . . .   . . .   you that . . .   ."
    [auxiliary verb]   [verb]   [sentence]
e.g., "I can assure you that it wasn't I."

I shall hold to the following terminological rules. An explicit speech action satisfies the standard form in its surface structure if it is made up of an *illocutionary* and a *propositional* component. The illocutionary component consists in an *illocutionary* act carried out with the aid of a *performative sentence*. This sentence is formed in the present indicative, affirmative, and has as its logical subject the first person and as its logical (direct) object the second person; the predicate, constructed with the help of a performative expression, permits in general the particle "hereby." [69] The performative component needs to be completed by a propositional component constructed by means of a sentence with *propositional content*. Whenever it is used in constative speech acts, the sentence with propositional content takes the form of a *propositional sentence [Aussagesatz]*. In its elementary form, the propositional sentence contains: (1) a name or a referring expression, with the aid of which the speaker identifies an object about which he wants to say something; and (2) a predicate expression for the general determination that the speaker wants to grant or deny to the object. In nonconstative speech acts, the propositional content is not stated but mentioned; in this case, propositional content coincides with what is usually called the unasserted proposition. (Thus I distinguish between the nominalized proposition "that p," which expresses a state of affairs, and the proposition "p," which represents a fact and which owes its assertoric force to the circumstance that it is embedded in a speech action of the type "assertion," and is thereby connected with an illocutionary act of asserting. In formal logic, of course, we treat propositions as autonomous units. Only the truth value we assign to "p" in contradistinction to "that p" is a reminder of the connection of the proposition with some constative speech act, a connection that is systematically ignored.) [70]

I shall call speech acts that have this structure *propositionally differentiated*. They are distinguished from symbolically mediated interactions—for example, a shout of "Fire!" that releases complementary actions, assistance, or flight—in that a propositional component of speech is uncoupled from the illocutionary act, so that (1) the propositional content can be held invariant across changes in illocutionary potential, and (2) the holistic mode of speech, in which representation, expression, and behavioral expectation are still one is replaced by differential modes of speech. I shall come back to this point in the following section. For the present, it suffices to point out that this level of differentiation of speech is a precondition for an action's ability to take on representational functions, that is, to say something about the world, either directly in the form of a statement or indirectly through mentioning a propositional content in nonconstative speech acts.

Explicit speech actions always have a propositional component in which a state of affairs is expressed. Nonlinguistic actions normally lack this component; thus they cannot assume representational functions. Signaling to a taxi so that I can begin working in my office at eight in the morning, reacting to the news of miserable marks with the desperate look of a father, joining a demonstration march, expressing nonacceptance of an invitation by not showing up, shaking a candidate's hand after he has passed the test, and so on and so forth, I observe or violate conventions. Naturally these normative expectations have a propositional content; but the propositional content must already be known to the participants if the expressed behavior is to be understandable as beginning work, a father's reaction, taking part in a demonstration—in short, as an action. The nonverbal expression itself cannot bring the propositional content of the presupposed norm to expression because it cannot take on representational functions. It can, of course, be understood as an indicator that calls to mind the propositional content of the presupposed norm.

Owing to their representational function, propositionally differentiated speech actions preserve for the actor more degrees of freedom in following norms. Beginning work at eight in the morning leaves only the option, to appear or not to appear; in the former case, to be on time or to be late; in the latter case, to be excused or unexcused, and so on. Nonverbal actions are often the result of such "trees" of yes/no decisions. But if the actor can express himself verbally, his situation is rich with alternatives. He can express the same speech act, say a command, in a very differentiated way; he will fulfill the same role segment, say that of a German teacher during class dictation, with very different speech acts. In short, propositionally differentiated speech leaves the actor more degrees of freedom in relation to a recognized normative background than does a nonlinguistic interaction.

Of course, propositionally differentiated utterances do not always have a linguistic form, as is shown by the example of grammaticalized sign language, for example, the standardized language of the deaf and dumb. In this connection, one might also mention pointing gestures, which represent an equivalent use of referential terms and thus supplement propositional speech. On the other hand, there are also speech actions that are not propositionally differentiated: namely, illocutionarily abbreviated speech actions, such as "Hello!" as a greeting formula, or "Check!" and "Checkmate!" as performative expressions for moves in a game and their consequences. The circumstance that a propositional component is lacking places the verbal utterances on a level with normal nonverbal actions; while such actions do refer to the propositional content of a presupposed convention, they do not express it.

As a first step in delimiting the pragmatic units of analysis, we can specify—out of the set of communicative actions that rest on the consensual foundation of reciprocally raised and recognized validity claims—the subset of *propositionally differentiated speech actions*. But this specification is not selective enough; for among these utterances we find such speech acts as "betting," "christening," "appointing," and so on. Despite their propositionally differentiated content, they are bound to a single institution (or to a narrowly circumscribed set of institutions); they can therefore be seen as the equivalent of actions that fulfill presupposed norms, either nonverbally or in an illocutionarily abbreviated way. The *institutional bond* of these speech acts can be seen in (among other things) the fact that the permissible propositional contents are narrowly limited by the normative meaning of betting, christening, appointing, marrying, and so on. One bets for stakes, christens with names, appoints to official positions, marries a partner, and so on. With institutionally bound speech actions, specific institutions are always involved. With institutionally unbound speech actions, only conditions of a generalized context must typically be met for a corresponding act to succeed. Institutionally bound speech actions express a specific institution in the same way that propositionally nondifferentiated and non-verbal actions express a presupposed norm. To explain what acts of betting or christening mean, I must refer to the institutions of betting or christening. By contrast, commands or advice or questions do not represent institutions but types of speech acts that can fit very different institutions. To be sure, "institutional bond" is a criterion that does not always permit an unambiguous classification. Commands can exist wherever relations of authority are institutionalized; appointments presuppose special, bureaucratically developed organizations; and marriages require a single institution (which is, however, found universally). But this does not destroy the usefulness of the analytic point of view. Institutionally unbound speech actions, insofar as they have any regulative meaning at all, are related to various aspects of action norms in general; they are not essentially fixed by particular institutions.

We can now define the desired analytic units as *propositionally differentiated* and *institutionally unbound speech actions*. Naturally, only those with an explicit linguistic form are suitable for analysis. Usually the context in which speech actions are embedded makes standard linguistic forms superfluous; for example, when the performative meaning is determined exclusively by the context of utterance; or when the performative meaning is only indicated, that is, expressed through inflection, punctuation, word position, or particles such as "isn't it?," "right?," "indeed," "clearly," "surely," and similar expressions.

Finally we shall exclude those explicit speech actions in standard form that appear in contexts that produce shifts of meaning. This is the case when the pragmatic meaning of a context-dependent speech act diverges from the meaning of the sentences used in it (and from that of the indicated conditions of a generalized context that have to be met for the type of speech action in question). Searle's "principle of expressibility" takes this into account: assuming that the speaker expresses his intention precisely, explicitly, and literally, it is possible in principle for every speech act carried out or capable of being carried out to be specified by a complex sentence.

Kanngiesser has given this principle the following form: "For every meaning x, it is the case that, if there is a speaker S in a language community P who means x, then it is possible that there be an expression in the language spoken by P which is an exact expression of x." [71] For our purposes, we can weaken this postulate to require that in a given language, for every interpersonal relation that a speaker wants to take up explicitly with another member of his language community, a suitable performative expression is either available or, if necessary, can be introduced through a specification of available expressions. With this modification, we can take into account reservations that have been expressed concerning Searle's principle.[72] In any case the heuristic meaning is clear—if the postulate of expressibility is valid, analysis can be limited to institutionally unbound, explicit speech actions in standard form.

The diagram sums up the viewpoints from which I have delimited the class of speech acts basic for analysis.

Derivation of the Analytic Units of the Theory of Speech Acts

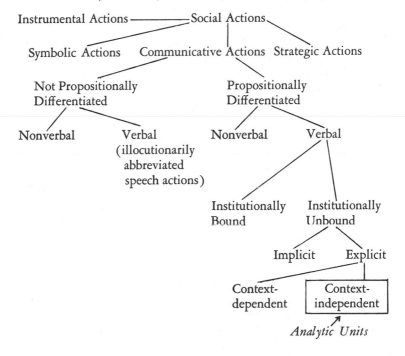

I have not explained the embedding of communicative action ("oriented to reaching understanding") in other types of action. It seems to me that *strategic action* ("oriented to the actor's success"—in general, modes of action that correspond to the utilitarian model of purposive-rational action) as well as (the still-insufficiently-analyzed) *symbolic action* (e.g., a concert, a dance —in general, modes of action that are bound to nonpropositional systems of symbolic expression) differ from communicative action in that individual validity claims are suspended (in strategic action, truthfulness, in symbolic action, truth).[73] My previous analyses of "labor" and "interaction" have not yet adequately captured the most general differentiating characteristics of instrumental and social (or communicative) action. I cannot pursue this desideratum here.

On the Double Structure of Speech

I would like to come back now to the characteristic *double structure* that can be read off the standard form of speech actions. Obviously the two components, the illocutionary and the propositional, can vary independently of one another. We can hold a propositional content invariant vis-à-vis the different types of speech acts in which it appears. In this abstraction of propositional content from the asserted proposition, a fundamental accomplishment of our language is expressed. Propositionally differentiated speech distinguishes itself therein from the symbolically mediated interaction we can already observe among primates.[74] Any number of examples of the *speech-act-invariance of propositional content* can be provided—for instance, for the propositional content "Peter's smoking a pipe" the following:

"I assert that Peter smokes a pipe."

"I beg of you (Peter) that you smoke a pipe."

"I ask you (Peter), do you smoke a pipe?"

"I warn you (Peter), smoke a pipe."

In a genetic perspective, the speech-act-invariance of propositional contents appears as an *uncoupling of illocutionary and propositional components* in the formation and transformation of speech actions. This uncoupling is a condition for the differentiation of the double structure of speech, that is, for the separation of two communicative levels on which speaker and hearer must simultaneously come to an understanding if they want to communicate their intentions to one another. I would distinguish (1) the *level of intersubjectivity* on which speaker and hearer, through illocutionary acts, establish the relations that permit them to come to an understanding with one another, and (2) the *level of propositional content* which is communicated. Corresponding to the relational and the content aspects in which every utterance can be analyzed, there are (in the standard form) the illocutionary and the propositional components of the speech act. The illocutionary act fixes the sense in which the propositional content is employed, and the act-complement determines the content that is understood "as something . . ." in the communicative function specified. (The hermeneutic "as" can be differentiated on both communicative levels. With a proposition "P" an identifiable object, whose existence is presupposed, can be characterized as something—e.g., as a "red," "soft," or "ideal" object. In connection with an illocutionary act, that is, through being embedded in a speech act, this propositional content can in turn be uttered as something—e.g., as a command or assertion.) A basic feature of language is connected with this double structure of speech, namely its inherent reflexivity. The standardized possibilities for directly and indirectly mentioning speech only make explicit a self-reference that is already contained in every speech act. In filling out the double structure of speech participants in dialogue communicate on two levels simultaneously. They combine communication of a content with communication about the role in which the communicated content is used. The expression *communication about* might be misleading here because it could be associated with *metalanguage* and would then bring to mind an idea of language levels according to which at every higher level metalinguistic statements about the object language of the next lower level can be made. But the concept of a hierarchy of language was introduced for formal languages, in which just that reflexivity of ordinary language is lacking. Moreover, in a metalanguage one always refers to an object language in the objectivating attitude of someone asserting facts or observing events; one forms metalinguistic *statements*. By contrast, on the level of intersubjectivity one chooses the illocutionary role in which the propositional content is to be used; and this communication about the sense in which the sentence with propositional content is to be employed requires a performative attitude on the part of those communicating. Thus the peculiar reflexivity of natural language rests in the first instance on the combination of a communication of content—effected in an objectivating attitude—with a communication concerning the relational aspect in which the content is to be understood—effected in a performative attitude.

Of course, participants in dialogue normally have the option of objectifying every illocutionary act performed as the content of another, a subsequent speech act. They can adopt an objectivating attitude toward the illocutionary component of a former speech act and shift this component to the level of propositional contents. Naturally they can do so only in performing a new speech act that has, in turn, a nonobjectified illocutionary component. The direct and indirect mention of speech standardizes this possibility of rendering explicit the reflexivity of natural language. The communication that takes place on the level of intersubjectivity in a speech act at t_n can be depicted on the level of propositional content in a further (constative) speech act at t_{n+1}. On the other hand, it is not possible simultaneously to perform and to objectify an illocutionary act.[75]

This option is sometimes the occasion for a descriptivist fallacy to which even pragmatic theories fall prey. We can analyze the structures of speech, as every other object, only in an objectivating attitude. In doing so, the actually accompanying illocutionary component cannot, as we saw above, become *uno acto* the object. This circumstance misleads many language theorists into the view that communication processes take place at a single level, namely that of transmitting content (i.e., information). In this perspective, the relational aspect loses its independence vis-à-vis the content aspect; the communicative role of an utterance loses its constitutive significance and is added to the information content. The pragmatic operator of the statement, which in formalized presentations (e.g., deontic logics) represents the illocutionary component of an utterance, is then no longer interpreted as a specific mode of reaching understanding about propositional contents but falsely as part of the information transmitted.

As opposed to this, I consider the task of universal pragmatics to be the rational reconstruction of the double structure of speech. Taking Austin's theory of speech acts as my point of departure, [in the next two sections] I would like now to make this task more precise in relation to the problems of meaning and validity.

Universal-Pragmatic Categories of Meaning

Austin's contrasting of locutionary and illocutionary acts set off a broad discussion that also brought some clarification to the theory of meaning. Austin reserved the concept *meaning* for the meaning of sentences with propositional content, while he used the concept *force* only for the illocutionary act of uttering sentences with propositional content. This leads to the following constellations:

meaning: sense and reference, locutionary act

force: attempt to reach an uptake, illocutionary act

Austin could point to the fact that sentences with the same propositional content could be uttered in speech acts of different types, that is, with differing illocutionary force. Nevertheless, the proposed distinction is unsatisfactory. If one introduces meaning only in a linguistic sense, as sentence meaning (in which either sentence meaning is conceived as a function of word meanings or, with Frege, word meanings are conceived as functions of possible sentence meanings), the restriction to the propositional components of speech acts is not plausible. Obviously their illocutionary components also have a meaning in a linguistic sense. In the case of an explicitly performative utterance, the performative verb employed has a lexical meaning, and the performative sentence constructed with its help has a meaning similar in a way to the dependent sentence with propositional content. "What Austin calls the illocutionary force of an utterance is that aspect

of its meaning which is either conveyed by its explicitly performative prefix, if it has one, or might have been so conveyed
by the use of such an expression." [76]

This argument neglects, however, the fact that force is something which, in a specific sense, belongs only to utterances and
not to sentences. Thus one might first hit upon the idea of reserving "force" for the meaning content that accrues to the sentence through its being uttered, that is, embedded in structures of
speech. We can certainly distinguish the phenomenon of meaning
that comes about through the employment of a sentence in an
utterance from mere sentence meaning. We can speak in a pragmatic sense of the meaning of an utterance, as we do in a linguistic
sense of the meaning of a sentence. Thus Alston has taken the
fact that the same speech acts can be performed with different
sentences as a reason for granting pragmatic meaning a certain
priority over linguistic meaning. In accordance with a consistent
use theory of meaning, he suggests that sentence (and word)
meanings are a function of the meaning of the speech acts in
which they are principally used. [77] The difficulty with this proposal is that it does not adequately take into account the relative
independence of sentence meanings in relation to the contingent
changes of meaning that a sentence can undergo when used in
different contexts. Moreover, the meaning of a sentence is apparently less dependent on the intention of the speaker than is
the meaning of an utterance.

Even if a sentence is very often used with different intentions
and in a context that pragmatically shifts meaning, its linguistic
meaning does not have to change. Thus, for example, when certain social roles prescribe that commands be uttered in the form
of requests, the pragmatic meaning of the utterance (as a command) in no way alters the linguistic meaning of the sentence
uttered (as a request). This is an additional reason for singling
out the standard conditions under which the pragmatic meaning
of an explicit speech action coincides with the linguistic meaning
of the sentences employed in it. Precisely in the case of an explicit
speech act in standard form, however, the categorial difference
between the meaning of expressions originally used in propositional sentences and the meaning of illocutionary forces (as well

as of expressed intentions) comes into view. This shows it does not make sense to explicate the concepts *meaning* versus *force* with reference to the distinction between the linguistic meaning of a sentence and the pragmatic meaning of an utterance.

The linguistic analysis of sentence meaning abstracts from certain relations to reality into which a sentence is put as soon as it is uttered and from the validity claims under which it is thereby placed. On the other hand, a consistent analysis of meaning is not possible without reference to some situations of possible use. Every linguistic expression can be used to form statements. Even illocutionary phrases (and originally intentional expressions) can become part of propositional sentences. This suggests that we might secure a certain unity for the linguistic analysis of the meanings of linguistic expressions by relating it in every case to their possible contribution to forming propositions. But that makes sense only for such expressions as can appear exclusively in propositional components of speech. The meaning of performative expressions should, on the contrary, be clarified by referring to their possible contribution to forming illocutionary acts (and the meaning of originally intentional expressions by referring to their possible contribution to forming first-person sentences expressing wishes, feelings, intentions, etc.). The linguistic explication of the meaning of "to promise" should orient itself around its contribution to forming the sentence,

> 1) "I hereby promise you that . . ."

and not,

> 2) "He is promising him that . . ."

Correspondingly, the explication of the meaning of "to hate" should refer to the sentence,

> 1') "I hate you."

instead of to the sentence,

> 2') "He hates him."

Only because and so long as the linguistic analysis of meaning is biased in favor of the propositionalized forms (2 or 2') is it necessary to supplement the meaning of propositional sentences with the meaning of the illocutionary force of an utterance (and the intention of a speaker). No doubt this circumstance motivated Austin to draw his distinction between meaning and force. To my mind, it would be better to differentiate the linguistic meanings of expressions according to their possible contribution to forming different types of speech acts (and different components of speech acts). Let us consider two examples:

3) "I'm notifying you that father's new car is yellow."

4) "I'm asking you, is father's new car yellow?"

Understanding the two (different) illocutionary acts is tied to other presuppositions than understanding their (concordant) propositional content. The difference becomes perceptible as soon as I go back to the conditions that must be fulfilled by situations in which someone who does not know English might learn (i.e., originally understand) the meanings. A hearer can understand the meaning of the sentence with the *propositional content:* "the being yellow of father's new car" under the condition that he has learned to use the propositional sentence correctly in the assertion:

5) "I'm telling you, father's new car is yellow,"

in order, for example, to report the observation that father's new car is yellow. The ability to have this or a similar experience must be presupposed. A proper use of the propositional sentence in (5) demands (at least) the following of the speaker:

a) The existence presupposition—there is one and only one object to which the characteristic "father's new car" applies.

b) The presupposition of identifiability—the (denotatively employed) propositional content contained in the characterization "father's new car" is a sufficient indication, in a given context, for a hearer to select the (and only the) object to which the characteristic applies.

c) The act of predication—the predicate "yellow" can be attributed to the object that is denoted.

Correspondingly, understanding the meaning of the propositional sentence contained in (5) demands of the hearer that he a′) Share-the-speaker presupposition (a).

b′) Fulfill-the-speaker presupposition (b), that is, actually identify the object referred to,

c′) Undertake for his part the act of predication (c).

It is otherwise with the *illocutionary components* of utterances (3) and (4). A hearer can understand the meaning of notifying or asking under the condition that he has learned to take part in successful speech acts of the following type:

6) "I (hereby) notify you that . . ."

7) "I (hereby) ask you whether . . ."

that is, has learned to assume both the role of the (acting) speaker as well as that of the (cooperating) hearer. The performance of an illocutionary act cannot serve to report an observation as the use of a propositional sentence can; and the ability to have perceptions is also not essentially presupposed here. Rather, conversely, the execution of a speech act is a condition of possibility of an experience, namely the communicative experience that the hearer has when he accepts the offer contained in the attempted speech action and enters into the requested connection with the speaker.

Whereas understanding (5) *presupposes* the possibility of sensory experiences (experiences of the type: observation), understanding (6) and (7) itself *represents* a communicative experience (an experience of the type: "participant observation").

The difference between originally illocutionary and originally propositional meanings (force and meaning in Austin's sense) can be traced back to differences in possible learning situations. We learn the meaning of illocutionary acts only in the performative attitude of participants in speech actions. By contrast, we learn the meaning of sentences with propositional content in the objectivating attitude of observers who correctly report their experiences in propositions.[78] We acquire originally illocutionary meanings in connection with communicative experiences that we have in entering upon the level of intersubjectivity and establishing interpersonal relations. We acquire originally propositional meanings in reporting something that happens in the world.

This difference notwithstanding, meanings acquired in a performative attitude can, of course, also appear in sentences with propositional content:

8) "I assure you that he notified me yesterday that . . ."

9) "I'm reporting to you that he asked me yesterday whether . . ."

This fact may explain why the indicated difference between the two categories of meaning is often not noticed. In sentences of propositional content, however, we can distinguish the meanings of expressions that can be used in a performative attitude from the word meanings that are permitted only as meaning components of sentences with propositional content. In utterances like (8) and (9), "notify" and "ask" bear a shade of meaning borrowed from the power that they have only in illocutionary roles—as in (6) and (7).

We can retain Austin's distinction between force and meaning in the sense of these two categories of meaning. Force then stands for the meaning of expressions that are originally used in connection with illocutionary acts, and meaning for the meaning of expressions originally used in connection with propositions. Thus we distinguish force and meaning as two categories of meaning that arise in regard to the general pragmatic functions of the *establishment of interpersonal relations,* on the one hand, and the *representation of facts,* on the other. (The third category of meaning, which corresponds to the function of *expression,* that is, to the disclosure of the speaker's wishes, feelings, intentions, etc. in first-person sentences, I shall leave to one side here, although reflections similar to those carried through for illocutionary acts apply to them as well.)

I would like to hold fast to the following results:

a. It is not advisable to reserve the concept of semantic content for the propositional component of a speech action and to characterize the meaning of the illocutionary component only by a pragmatic operator (which designates a specific illocutionary force).

b. On the other hand, it is also unsatisfactory to reconstruct the meaning of a performative sentence in exactly the same way as the meaning of a sentence with propositional content; the illocutionary component of a speech act neither expresses a proposition nor mentions a propositional content.[79]

c. It is equally unsatisfactory to equate illocutionary force with the meaning components that are added to the meaning of a sentence through the act of utterance.

d. From a universal-pragmatic point of view, the meanings of linguistic expressions can be categorically distinguished according to whether they can appear only in sentences that take on a representational function or whether they can specifically serve to establish interpersonal relations or to express intentions.[80]

Thematization of Validity Claims and Modes of Communication

Austin's contrast of locutionary and illocutionary acts has become important not only for the theory of meaning; the discussions about basic types of speech action and basic modes of language has also fastened on to this pair of concepts. At first Austin wanted to draw the boundary in such a way that "the performative should be doing something as opposed to just saying something; and the performative is happy or unhappy as opposed to true and false." [81] From this there results the following correlations:

locutionary acts: constatives, true/untrue.

illocutionary acts: performatives, happy/unhappy.

But this demarcation of locutionary and illocutionary acts could not be maintained when it became apparent that all speech actions—the constative included—contain a locutionary component (in the form of a sentence with propositional content) and an illocutionary component (in the form of a performative sentence).[82] What Austin had first introduced as the locutionary act was now replaced by (a) the propositional component contained in every explicit performative utterance, and (b) a special class of illocutionary acts that imply the validity claim of truth—constative speech acts. Austin himself later regarded constative speech acts as only one of the different classes of speech acts. The two sentences:

1) "I assert that . . ."

2) "I'm warning you that . . ."

equally express illocutionary acts.[83] But this has the interesting
consequence that the validity claim contained in constative speech
acts (truth/falsity) represents only a special case among the
validity claims that speakers, in speech acts, raise and offer for
vindication vis-à-vis hearers.

In general we may say this: with both statements (and, for example,
descriptions) and warnings, etc., the question of whether, granting
that you did warn and had the right to warn, did state or did advise,
you were *right* to state or to warn or advise, can arise—not in the sense
of whether it was opportune or expedient, but whether, on the facts
and your knowledge of the facts and the purpose for which you were
speaking, and so on, this was the proper thing to say.[84]

In this passage Austin emphasizes the claims to be right, or
validity claims, that we raise with any (and not just with constative) speech acts. But he distinguishes these only incidentally
from the conditions of the generalized context that typically must
be fulfilled if a speech act of the corresponding type is to succeed
(that is, from happiness/unhappiness conditions in general). It
is true of assertions, in the same way as it is of warnings, advis-
ings, promisings, and so forth, that they can succeed only if both
conditions are fulfilled: (a) to be in order, and (b) to be right.

But the real conclusion must surely be that we need . . . to establish
with respect to each kind of illocutionary act—warnings, estimates,
verdicts, statements, and descriptions—what if any is the specific way
in which they are intended, first to be in order or not in order, and
second, to be "right" or "wrong"; what terms of appraisal and dis-
appraisal are used for each and what they mean. This is a wide field
and certainly will not lead to a simple distinction of true and false;
nor will it lead to a distinction of statements from the rest, for stating
is only one among very numerous speech acts of the illocutionary
class.[85]

Speech acts can be in order with respect to typically restricted
contexts (a); but they can be valid only with respect to the

fundamental claim that the speaker raises with his illocutionary act (b). I shall be coming back to both of these classes of conditions that must be fulfilled in order for speech acts to succeed. At this point I am interested only in the fact that the comparison between constative and nonconstative speech acts throws light on the validity basis that manifestly underlies *all* speech actions.

To be sure, this does bring out the special position of constative speech acts. Assertions do not differ from other types of speech actions in their performative/propositional double structure, nor by virtue of conditions of a generalized context, for these vary in a typical way for all speech actions; but they do differ from (almost) all other types of speech actions in that they prima facie imply an unmistakable validity claim, a truth claim. It is undeniable that other types of speech actions also imply *some* validity claim; but in determining exactly what validity claim they imply, we seldom encounter so clear and universally recognized a validity claim as "truth" (in the sense of propositional truth). It is easy to see the reason for this; the validity claim of constative speech acts is presupposed in a certain way by speech acts of *every* type. The meaning of the propositional content mentioned in nonconstative speech acts can be made explicit through transforming a sentence of propositional content, "that p," into a propositional sentence "p"; and the truth claim belongs essentially to the meaning of the proposition thereby expressed. Truth claims are thus a type of validity claim built into the structure of possible speech in general. Truth is a universal validity claim; its universality is reflected in the double structure of speech.

Looking back, Austin assures himself of what he originally had in mind with his contrast of constative and nonconstative speech actions (constatives versus performatives):

With the constative utterances, we abstract from the illocutionary ... aspects of the speech act, and we concentrate on the locutionary; moreover, we use an oversimplified notion of correspondence with the facts ... We aim at the ideal of what would be right to say in all circumstances, for any purpose, to any audience, etc. Perhaps this is sometimes realized. With the performative we attend as much as possible to the illocutionary force of the utterance, and abstract from the dimension of correspondence with facts.[86]

After he recognized that constative speech acts represent only one of several types of speech action, Austin gave up the aforementioned contrast in favor of a set of unordered families of speech actions. I am of the opinion, however, that what he intended with the contrast constative versus performative can be adequately reconstructed.

We have seen that communication in language can take place only when the participants, in communicating with one another about something, simultaneously enter upon two levels of communication—the level of intersubjectivity on which they take up interpersonal relations and the level of propositional contents. However, in speaking we can make either the interpersonal relation or the propositional content more centrally thematic; correspondingly we make a more interactive or a more cognitive use of our language. In the *interactive use of language,* we thematize the relations into which speaker and hearer enter—as a warning, promise, request—while we only mention the propositional content of the utterances. In the *cognitive use of language,* by contrast, we thematize the content of the utterance as a proposition about something that is happening in the world (or that could be the case), while we only indirectly express the interpersonal relation. This incidental character can be seen, for example, in the fact that in English the explicit form of assertion ("I am asserting (to you) that . . ."), although grammatically correct, is rare in comparison to the short form that disregards the interpersonal relation.

As the content is thematized in the cognitive use of language, only speech acts in which propositional contents assume the explicit form of propositions are permitted here. With these constative speech acts, we raise a truth claim for the proposition asserted. In the interactive use of language, in which the interpersonal relation is thematically stressed, we refer in various ways to the validity of the normative context of the speech action.

For this latter use the (authorized) command has a paradigmatic significance similar to that of the assertion for the cognitive use of language. Truth is merely the most conspicuous—not the only—validity claim reflected in the formal structures of speech. The illocutionary force of the speech act, which produces a legitimate (or illegitimate) interpersonal relation between the participants, is borrowed from the binding force of recognized norms of action (or of evaluation); to the extent that a speech act is an action, it actualizes an already-established pattern of relations. The validity of a normative background of institutions, roles, socioculturally habitual forms of life—that is, of conventions—is always presupposed. This by no means holds true only for institutionally bound speech actions such as betting, greeting, christening, appointing, and the like, each of which satisfies a *specific* institution (or a narrowly circumscribed class of norms). In promises too, in recommendations, prohibitions, prescriptions, and the like, which are not regulated from the outset by institutions, the speaker implies a validity claim that must, if the speech acts are to succeed, be covered by existing norms, and that means by (at least) de facto recognition of *the claim that these norms rightfully exist.* This internal relation between the validity claims implicitly raised in speech actions and the validity of their normative context is stressed in the interactive use of language, as is the truth claim in the cognitive use of language. Just as only constative speech acts are permitted in the cognitive use of language, so in the interactive use only those speech acts are permitted that characterize a specific relation that speaker and hearer can adopt to norms of action or evaluation. I call these regulative speech acts.[87] With the illocutionary force of speech actions, the normative validity claim—rightness or appropriateness [*Richtigkeit, Angemessenheit*]—is built just as universally into the structures of speech as the truth claim. But the validity claim of a normative context is explicitly invoked only in regulative speech acts (in commands and admonitions, in prohibitions and refusals, in promises and agreements, notices, excuses, recommendations, admissions, and so forth). The truth reference of the mentioned propositional content remains, by contrast, merely implicit; it pertains only to its existential presuppositions. Conversely, in constative speech acts, which explicitly raise a truth claim, the normative validity claim remains implicit, although these too (e.g., reports, explications, communications, elucidations, narrations, and so forth) must correspond to an established pattern of value orientations—that is, they must fit a recognized normative context—if the interpersonal relations intended in them are to come to pass.

It seems to me that what Austin had in mind with his (later abandoned) classification into constative versus performative utterances is captured in the distinction between the cognitive and the interactive uses of language. In the cognitive use of language, with the help of constative speech acts, we thematize the propositional content of an utterance; in the interactive use of language, with the help of regulative speech acts, we thematize the kind of interpersonal relation established. The difference in thematization results from stressing one of the validity claims universally inhabiting speech, that is, from the fact that in the cognitive use of language we raise truth claims for propositions and in the interactive use of language we claim (or contest) the validity of a normative context for interpersonal relations. Austin himself did not draw this consequence because, on the one hand, he took only one universal validity claim into consideration, namely, propositional truth interpreted in terms of the correspondence theory of truth; but he wanted, on the other hand, to make this single validity claim compatible with many types of speech acts other than constative speech acts. In his words: "If, then, we loosen up our ideas of truth and falsity we shall see that statements, when assessed in relation to the facts, are not so different after all from pieces of advice, warnings, verdicts and so on." [88] This loosening of the concept of truth in favor of a broad dimension of evaluation, in which an assertion can be just as well characterized as exaggerated or precise or inappropriate as true or false, results somehow in the assimilation of all validity claims to that of propositional truth. "We see that, when we have an order or a warning or a piece of advice, there is a question about how this is related to fact which is not perhaps so different from the kind of question that arises when we discuss how a statement is related to fact." [89] It seems to me that Austin confuses the validity claim of propositional truth, which can be understood in the first instance in terms of a correspondence between statements and facts, with the validity claim of normative rightness, which does not fit the correspondence theory by truth.

To the extent that warnings or pieces of advice rest on predictions, they are part of a cognitive use of speech. Whether those involved were right to utter certain warnings or pieces of advice in a given situation, depends in this case on the truth of the corresponding predictions. As part of an interactive use of speech, warnings and pieces of advice can also have a normative meaning. Then the right to issue certain warnings and advice depends on whether the presupposed norms to which they refer are valid (that is, are intersubjectively recognized) or not (and, at a next stage, ought or ought not to be valid). But most types of speech action can be more clearly attached to a single mode of language use. Whether an estimate is good or bad clearly depends on the truth of a corresponding statement; estimates usually appear in the cognitive use of language. On the other hand, whether the verdict of a court, the reprimand of a person, or the command of a superior to a subordinate with regard to certain behavior are justly pronounced, deservedly issued, or rightfully given depends just as clearly on whether a recognized norm has been correctly applied to a given case (or whether the right norm has been applied to the case). Legal verdicts, reprimands, and orders can only be part of an interactive use of language. Austin himself once considered the objection that different validity claims are at work in these cases:

Allowing that, in declaring the accused guilty, you have reached your verdict properly and in good faith, it still remains to ask whether the verdict was just, or fair. Allowing that you had the right to reprimand him as you did, and that you have acted without malice, one can still ask whether your reprimand was deserved . . . There is one thing that people will be particularly tempted to bring up as an objection against any comparison between this second kind of criticism and the kind appropriate to statements, and that is this: aren't these questions about something's being good, or just, or fair, or deserved entirely distinct from questions of truth and falsehood? That, surely, is a very simple black-and-white business: either the utterance corresponds to the facts or it doesn't, and that's that.[90]

In comprehending the universal validity claim of truth in the same class with a host of particular evaluative criteria, Austin obliterated the distinction between the clear-cut universal validity claims of propositional truth and normative rightness (and truthfulness). But this proves to be unnecessary if in a given speech action we distinguish among

    a. The implicitly presupposed conditions of generalized context,
    b. The specific meaning of an interpersonal relation to be established, and
    c. The implicitly raised, general validity claim.

Whereas (a) and (b) fix the distinct classes (different in different languages) of standardized speech actions, (c) determines the universal modes of communication, modes inherent in speech in general.

Before going into (a) and (b), I would like at least to remark that the Austinian starting point in the distinction between performative and constative utterances provides an overly narrow view; the validity spectrum of speech is not exhausted by the two modes of communication I developed from this distinction. Naturally there can be no mode of communication in which the intelligibility of an utterance is thematically stressed; for every speech act must fulfill the presupposition of *comprehensibility* in the same way. If in some communication there is a breakdown of intelligibility, the requirement of comprehensibility can be made thematic only through passing over to a hermeneutic discourse, and then in connection with the relevant linguistic system. The *truthfulness* with which a speaker utters his intentions can, however, be stressed at the level of communicative action in the same way as the truth of a proposition and the rightness (or appropriateness) of an interpersonal relation. Truthfulness guarantees the transparency of a subjectivity representing itself in language. It is especially emphasized in *the expressive use of language*. The paradigms are first-person sentences in which the speaker's wishes, feelings, intentions, etc. (which are expressed incidentally in every speech act) become disclosed, that is, sentences such as:

    3) "I long for you."

    4) "I wish that . . ."

It is unusual for such sentences to be explicitly embedded in an illocutionary act as follows:

    3') "I hereby express to you that I long for you."

In the expressive use of language the interpersonal relation carrying the function of public self-representation is not thematic and thus need be mentioned only in situations in which the presupposition of the speaker's truthfulness is not taken for granted; for this, avowals are the paradigm:

> 5) "I must confess to you that . . ."
> 6) "I don't want to conceal from you that . . ."

For this reason, expressive speech acts such as disclosing, concealing, revealing, and the like, cannot be correlated with the expressive use of language (which can, in a way, dispense with illocutionary acts) in the same manner as constative speech acts are correlated with the cognitive use of language and regulative speech acts with the interactive. Nevertheless, truthfulness too is a universal impliction of speech, as long as the presuppositions of communicative action are not altogether suspended. In the cognitive use of language the speaker must, in a trival sense, truthfully express his thoughts, opinions, assumptions, and so forth; since he asserts a proposition, however, what matters is not the truthfulness of his intentions, but the truth of the proposition. Similarly, in the interactive use of language the speaker expresses the intention of promising, reprimanding, refusing, and so forth; but since he brings about an interpersonal relation with a hearer, the truthfulness of his intention is only a necessary condition, whereas what is important is that the action fit a recognized normative context.

Thus we have the following correlations:

Mode of Communication	Type of Speech Action	Theme	Thematic Validity Claim
Cognitive	Constatives	Propositional content	Truth
Interactive	Regulatives	Interpersonal relation	Rightness, appropriateness
Expressive	Avowals	Speaker's intention	Truthfulness

(*P.S.*: The modes of language use can be demarcated from one another only paradigmatically. I am not claiming that every sequence of speech actions can be unequivocally classified under these viewpoints. I am claiming only that every competent speaker has in principle the possibility of unequivocally selecting one mode because with every speech act he *must* raise three universal validity claims, so that he *can* single out one of them to thematize a component of speech.)

The Rational Foundation of Illocutionary Force

Having somewhat elucidated the meaning structure and validity basis of basic types of speech acts, I would like to return to the question, in what does the illocutionary force of an utterance consist? To begin, we know only what it results in if the speech action succeeds—in bringing about an interpersonal relation. Austin and Searle have analyzed illocutionary force by looking for conditions of success or failure of speech acts. An uttered content receives a specific communicative function through the fact that the standard conditions for the occurrence of a corresponding interpersonal relation are fulfilled. With the illocutionary act, the speaker makes an offer that can be accepted or rejected. The attempt a speaker makes with a illocutionary act may founder for contingent reasons on the refusal of the addressee to enter into the proffered relationship. This case is of no interest in the present context. We shall be concerned with the other case, in which the speaker himself is responsible for the failure of the speech act because the utterance is unacceptable. When the speaker makes an utterance that manifestly contains no serious offer, he cannot count on the occurrence of the relationship intended by him.

I shall speak of the success of a speech act only when the hearer not only understands the meaning of the sentence uttered but also actually enters into the relationship intended by the speaker. And I shall analyze the conditions for the success of speech acts in terms of their "acceptability." Since I have restricted my examination from the outset to communicative action—that is, action oriented to reaching understanding—a speech act counts as acceptable only if the speaker not merely feigns but sincerely makes a serious offer. A serious offer demands a certain engagement on the part of the speaker. But before going into this, I would like to mention additional reasons for the unacceptability of illocutionary acts.

Austin developed his doctrine of "infelicities" primarily with a view to institutionally bound speech acts; for this reason, the examples of "misfires" (that is, of misinvocations, misexecutions, misapplications) are [viewed as] typical for all possible cases of rule violation. Thus the unacceptability of speech acts can stem from transgressions of underlying norms of action. If in a wedding ceremony a priest recites the prescribed marriage formula incorrectly or not at all, the mistake lies at the same level as, let us say, the command of a university lecturer in class to one of his students, who can reply to him (with right, let us assume): "You can indeed request a favor of me, but you cannot command me." The conditions of acceptability are not fulfilled; but in both cases, these conditions are defined by a given normative context. We are looking, by contrast, for conditions of acceptability that lie within the institutionally unbound speech act itself.

Searle analyzed the conventional presuppositions of different types of speech actions that must be fulfilled if their illocutionary force is to be comprehensible and acceptable. Under the title "preparatory rules," he specifies generalized or restricted *contexts* for possible types of speech actions. A promise, for example, is not acceptable if the following conditions, among other, are not fulfilled: (a) H (the hearer) prefers S's (the speaker's) doing A (a specific action) to his not doing A, and S moreover believes this to be the case; (b) it is not obvious to both S and H that S will do A in the normal course of events.[91] If conventional presuppositions of this kind are not fulfilled, the act of promising is pointless, that is, the attempt by a speaker to carry out the illocutionary act anyway makes no sense and is condemned to failure from the outset.[92]

The generalized context conditions for institutionally unbound speech actions are to be distinguished from the conditions for applying established norms of action.[93] The two sets of conditions of application, those for types of speech action and those for norms of action, must vary (largely) independently of one another if (institutionally unbound) speech actions are to represent a repertory from which the acting subject, with the help of a finite number of types, can put together any number of norm-conformative actions.

The peculiar force of the illocutionary—which in the case of institutionally unbound speech actions is not borrowed directly from the validity of established norms of action—cannot be explained by means of the speech-act-typical context restrictions. This is possible only with the help of the specific presuppositions that Searle introduces under the title "essential rules." In doing so, he appears, it is true, to succeed only in paraphrasing the meaning of the corresponding performative verbs (for example, requests: "count as an attempt to get H to do A"; or questions: "count as an attempt to elicit information from H"). It is interesting, however, that these circumscriptions include the common determination, "count as an attempt . . ." The essential presupposition for the success of an illocutionary act consists in the speaker's entering into a specific *engagement,* so that the hearer can rely on him. An utterance can count as a promise, assertion, request, question, or avowal, if and only if the speaker makes an offer that he is ready to make good insofar as it is accepted by the hearer. The speaker must engage himself, that is, indicate that in certain situations he will draw certain consequences for action. The *content* of the engagement[94] is to be distinguished from the *sincerity* of the engagement. This condition, introduced by Searle as the "sincerity rule," must *always* be fulfilled in the case of communicative action that is oriented to reaching understanding. Thus in what follows I shall, in speaking of the speaker's engagement, presuppose both a certain content of engagement and the sincerity with which the speaker is willing to enter into his engagement. So far as I can see, previous analyses of speech acts have been unsatisfactory, as they have not clarified the engagement of the speaker, on which the acceptability of his utterance specifically depends.

The discernible and sincere readiness of the speaker to enter into a specific kind of interpersonal bond has, compared with the general context conditions, a peculiar status. The restricted contexts that specific types of speech actions presuppose must (a) be given, and (b) be supposed to exist by the participants. Thus the following two statements must hold: (a) a statement to the effect that certain contexts obtain, indeed those required by the type in question; and (b) a statement to the effect that speaker and hearer suppose these contexts to obtain. The *specific* presupposition of speaker engagement, on the other hand, should not be analyzed in the same way, that is, so as to yield the following two statements: (a) a statement to the effect that there is a certain engagement on the part of the speaker; and (b) a statement to the effect that the hearer supposes this speaker engagement to obtain. One *could* choose this strategy of analysis; but I regard it as unsuitable. It would suggest that we speak of the existence of an engagement in the same sense as we speak of the existence of restricted contexts. I can ascertain in an appropriate manner, through observation or questioning, whether or not conditions of generalized contexts obtain; on the other hand, I can only *test* whether a speaker engages himself in a specific way and commits himself to certain consequences for action; I can ascertain at best whether there are sufficient indicators for the conjecture that the offer would withstand testing.

The bond into which the speaker is willing to enter with the performance of an illocutionary act means a guarantee that, in consequence of his utterance, he will fulfill certain conditions— for example, regard a question as settled when a satisfactory answer is given; drop an assertion when it proves to be false; follow his own advice when he finds himself in the same situation as the hearer; stress a request when it is not complied with; act in accordance with an intention disclosed by avowal, and so on. *Thus the illocutionary force of an acceptable speech act consists in the fact that it can move a hearer to rely on the speech-act-typical commitments of the speaker.* But if illocutionary force has more than a suggestive influence, what can motivate the hearer to base his action on the premise that the speaker seriously intends the engagement he indicates? When it is a question of institutionally bound speech actions, he can perhaps rely on the binding force of an established norm of action. In the case of institutionally unbound speech acts, however, illocutionary force cannot be traced back directly to the binding force of the normative context. The illocutionary force with which the speaker, in carrying out his speech act, influences the hearer, can be understood only if we take into consideration sequences of speech actions that are connected with one another on the basis of a reciprocal recognition of validity claims.

With their illocutionary acts, speaker and hearer raise validity claims and demand they be recognized. But this recognition need not follow irrationally, since the validity claims have a cognitive character and can be checked. I would like, therefore, to defend the following thesis: *In the final analysis, the speaker can illocutionarily influence the hearer and vice versa, because speech-act-typical commitments are connected with cognitively testable validity claims*—that is, because the reciprocal bonds have a rational basis. The engaged speaker normally connects the specific sense in which he would like to take up an interpersonal relationship with a thematically stressed validity claim and thereby chooses a specific mode of communication.

Thus assertions, descriptions, classifications, estimates, predictions, objections, and the like, have different specific meanings; but the claim put forward in these different interpersonal relationships is, or is based on, the truth of corresponding propositions or on the ability of a subject to have cognitions. Correspondingly, requests, orders, admonitions, promises, agreements, excuses, admissions, and the like, have different specific meanings; but the claim put forward in these different interpersonal relationships is, or refers to, the rightness of norms or to the ability of a subject to assume responsibility. We might say that in different speech acts the content of the speaker's engagement is *determined by different ways of appealing to the same, thematically stressed, universal validity claim.* And since as a result of this appeal to universal validity claims, the speech-act-typical commitments take on the character of obligations to provide grounds or to prove trustworthy, the hearer can be rationally motivated by the speaker's signaled engagement to accept the latter's offer. I would like to elucidate this for each of the three modes of communication.

In the cognitive use of language, the speaker proffers a speech-act-immanent *obligation to provide grounds [Bergründungsverpflichtung]*. Constative speech acts contain the offer to recur if necessary to the *experiential source* from which the speaker draws the *certainty* that his statement is true. If this immediate grounding does not dispel an ad hoc doubt, the persistingly problematic truth claim can become the subject of a theoretical discourse. In the interactive use of language, the speaker proffers a speech-act-immanent *obligation to provide justification* [*Rechtfertigungs-verpflichtung*]. Of course, regulative speech acts contain only the offer to indicate, if necessary, the *normative context* that gives the speaker the *conviction* that his utterance is right. Again, if this immediate justification does not dispel an ad hoc doubt, we can pass over to the level of discourse, in this case of practical discourse. In such a discourse, however, the subject of discursive examination is not the rightness claim directly connected with the speech act, but the validity claim of the underlying norm. Finally, in the expressive use of language the speaker also enters into a speech-act-immanent obligation, namely the *obligation to prove trustworthy* [*Bewährungsverpflichtung*], to show in the consequences of his action that he has expressed just that intention which actually guides his behavior. In case the immediate *assurance* expressing what is *evident* to the speaker himself cannot dispel ad hoc doubts, the truthfulness of the utterance can only be checked against the consistency of his subsequent behavior.

Every speech-act-immanent obligation can be made good at two levels, namely immediately, in the context of utterance— whether through recourse to an experiential base, through indicating a corresponding normative context, or through affirmation of what is evident to oneself—and mediately, in discourse or in subsequent actions. But only in the case of the obligations to ground and to prove trustworthy, into which we enter with constative and with expressive speech acts, do we refer to the *same* truth or truthfulness claim. The obligation to justify, into which we enter with regulative speech acts, refers immediately to the claim that the speech action performed fits an existing normative background; whereas with the entrance into practical discourse the topic of discussion is the validity of the very norm from which the rightness claim of the speaker is merely borrowed.

Our reflections have led to the following provisional results:

    a. A speech act succeeds, that is, it brings about the interpersonal relation that S intends with it, if it is

comprehensible and acceptable, and

accepted by the hearer.

    b. The acceptability of a speech act depends on (among other things) the fulfillment of two pragmatic presuppositions:

the existence of speech-act-typically restricted contexts (preparatory rule); and

a recognizable engagement of the speaker to enter into certain speech-act-typical obligations (essential rule, sincerity rule).

    c. The illocutionary force of a speech act consists in its capacity to move a hearer to act under the premise that the engagement signalled by the speaker is seriously meant:

in the case of institutionally bound speech acts, the speaker can borrow this force from the binding force of existing norms;

in the case of institutionally unbound speech acts, the speaker can develop this force by inducing the recognition of validity claims.

    d. Speaker and hearer can reciprocally motivate one another to recognize validity claims because the content of the speaker's engagement is determined by a specific reference to a thematically stressed validity claim, whereby the speaker, in a cognitively testable way, assumes

with a truth claim, obligations to provide grounds,

with a rightness claim, obligations to provide justification, and

with a truthfulness claim, obligations to prove trustworthy.

A Model of Linguistic Communication

The analysis of what Austin called the illocutionary force of an utterance has led us back to the validity basis of speech. Institutionally unbound speech acts owe their illocutionary force to a cluster of validity claims that speakers and hearers have to raise and recognize as justified if grammatical (and thus comprehensible) sentences are to be employed in such a way as to result in successful communication. A participant in communication acts with an orientation to reaching understanding only under the condition that, in employing comprehensible sentences in his speech acts, he raises three validity claims in an acceptable way. He claims truth for a stated propositional content or for the existential presuppositions of a mentioned propositional content. He claims rightness (or appropriateness) for norms (or values), which, in a given context, justify an interpersonal relation that is to be performatively established. Finally, he claims truthfulness for the intentions expressed. Of course, individual validity claims can be thematically stressed, whereby the truth of the propositional content comes to the fore in the cognitive use of language, the rightness (or appropriateness) of the interpersonal relation in the interactive, and the truthfulness of the speaker in the expressive. But in every instance of communicative action the system of all validity claims comes into play; they must always be raised simultaneously, although they cannot all be thematic at the same time.

The universality of the validity claims inherent in the structure of speech can perhaps be elucidated with reference to the systematic place of language. Language is the medium through which speakers and hearers realize certain fundamental demarcations. The subject demarcates himself: (1) from an environment that he objectifies in the third-person attitude of an observer; (2) from an environment that he conforms to or deviates from in the ego-alter attitude of a participant; (3) from his own subjectivity that he expresses or conceals in a first-person attitude; and finally (4) from the medium of language itself. For these domains of reality I have proposed the somewhat arbitrarily chosen terms: *external nature, society, internal nature,* and *language.* The validity claims unavoidably implied in speech oriented to reaching understanding show that these four regions must always simultaneously appear. I shall characterize the way in which these regions appear with a few phenomenological indications.

By *external nature* I mean the objectivated segment of reality that the adult subject is able (even if only mediately) to perceive and manipulate. One can, of course, adopt an objectivating attitude not only toward inanimate nature but toward all objects and states of affairs that are directly or indirectly accessible to sensory experience. *Society* designates that symbolically prestructured segment of reality that the adult subject can understand in a nonconformative attitude, that is, as one acting communicatively (as a participant in a system of communication). Legitimate interpersonal relations belong here, as do institutions, traditions, cultural values, etc. We can replace this conformative attitude with an objectivating attitude toward society; conversely, we can switch to a conformative attitude in domains in which (today) we normally behave objectivatingly—for example, in relation to animals and plants. I class as *internal nature* all wishes, feelings, intentions, etc., to which an "I" has privileged access and can express as its own experiences before a public. It is "precisely in this expressive attitude that the "I" knows itself not only as subjectivity but also as something that has always already transcended the bounds of mere subjectivity, in cognition, language, and interaction simultaneously. To be sure, if the subject adopts an objectivating attitude toward himself, this alters the sense in which intentions can be expressed.[95]

Finally, I introduced the linguistic medium of our utterances as a special region; precisely because language (including non-propositional symbol systems) remains in a peculiar half-transcendence in the performance of our communicative actions and expressions, it presents itself to the speaker and the actor (preconsciously) as a segment of reality *sui generis*. Again, this does not preclude our being able to adopt, in regard to linguistic utterances or systems of symbols, an objectivating attitude directed to the sounds or signs.

The model intuitively introduced here is that of a communication in which grammatical sentences are embedded, by way of universal validity claims, in three relations to reality, thereby assuming the corresponding pragmatic functions of representing facts, establishing legitimate interpersonal relations, and expressing one's own subjectivity. According to this model, language can be conceived as the medium of interrelating three worlds; for every successful communicative action there exists a threefold relation between the utterance and (a) "the external world" as the totality of existing states of affairs, (b) "our social world" as the totality of all normatively regulated interpersonal relations that count as legitimate in a given society, and (c) "a particular inner world" (of the speaker) as the totality of his intentional experiences.

We can examine every utterance to see whether it is true or untrue, justified or unjustified, truthful or untruthful, because in speech, no matter what the emphasis, grammatical sentences are embedded in relations to reality in such a way that in an acceptable speech action segments of external nature, society, and internal nature always come into appearance together. Language itself also appears in speech, for speech is a medium in which the linguistic means that are employed instrumentally are also reflected. In speech, speech sets itself off from the regions of external nature, society, and internal nature, as a reality sui generis, as soon as the sign-substrate, meaning, and denotation of a linguistic utterance can be distinguished.

The following table represents the correlations that obtain for

a. The domains of reality to which every speech action takes up relation.

b. The attitudes of the speaker prevailing in particular modes of communication.

c. The validity claims under which the relations to reality are established.

d. The general functions that grammatical sentences assume in their relations to reality.

Domains of Reality	Modes of Communication: Basic Attitudes	Validity Claims	General Functions of Speech
"The" World of External Nature	Cognitive: Ojectivating Attitude	Truth	Representation of Facts
"Our" World of Society	Interactive: Conformative Attitude	Rightness	Establishment of Legitimate Interpersonal Relations
"My" World of Internal Nature	Expressive: Expressive Attitude	Truthfulness	Disclosure of Speaker's Subjectivity
Language	———	Comprehensibility	———

2   Moral Development
and Ego Identity

In July of 1974, on the occasion of its fiftieth anniversary, the *Institut
für Sozialforschung* in Frankfurt arranged a series of lectures to which
Herbert Marcuse, Leo Löwenthal, Oskar Negt, Alfred Schmidt, and
Jürgen Habermas contributed. This is the text on which Habermas' lecture was based.

Since the tradition of the Frankfurt Institute has been immediately embodied in the lectures by Marcuse and Löwenthal and has
been made present in two essential aspects by contributions from
representatives of the postwar generation, I feel myself absolved
from duties that the occasion of this anniversary would otherwise have imposed. In other words, I shall not be delivering a
ceremonial address. Moreover, the state in which critical social
theory finds itself today—if one compares it with its now classical expressions—gives no occasion to celebrate. Finally, there is
a systematic reason for being somewhat sparing with tributes to
the past: the members of the original institute have always felt
themselves one with psychoanalysis in the intention of breaking
the power of the past over the present; to be sure, they have tried
to realize this intention, as psychoanalysis does, through future-
oriented memory.

I

I would like today to deal with fragments of a thematic that interests my co-workers and me in connection with an empirical investigation into the potential for conflict and apathy among young people.[1] We suspect that there is a connection between patterns of socialization, typical developments of adolescence, corresponding solutions to the adolescent crisis, and the forms of identity constructed by the young—a connection that can explain deep-seated, politically relevant attitudes. This problem leads one to reflect on moral development and ego identity, a theme that takes us naturally beyond this to a fundamental question of critical social theory, viz. to the question of the normative implications of its fundamental concepts. The concept of ego identity obviously has more than a descriptive meaning. It describes a symbolic organization of the ego that lays claim, on the one hand, to being a universal ideal, since it is found in the structures of formative processes in general and makes possible optimal solutions to culturally invariant, recurring problems of action. On the other hand, an autonomous ego organization is by no means a regular occurrence, the result, say, of naturelike processes of maturation; in fact it is usually not attained.

If one considers the normative implications of concepts such as ego strength, dismantling the ego-distant parts of the superego, and reducing the domain in which unconscious defense mechanisms function, it becomes clear that psychoanalysis also singles out certain personality structures as ideal. When psychoanalysis is interpreted as a form of language analysis, its normative meaning is exhibited in the fact that the structural model of ego, id, and superego presupposes unconstrained, pathologically undistorted communication.[2] In psychoanalytic literature these normative implications are, of course, usually rendered explicit in connection with the therapeutic goals of analytic treatment. In the social-psychological works of the *Institut für Sozialforschung* one can show that the basic concepts of psychoanalytic theory could enter integrally into description, hypothesis formation, and measuring instruments precisely because of their normative content. The early studies by Fromm of the sado-masochistic character and by Horkheimer of authority and the family, Adorno's investigation of the mechanisms for the formation of prejudice in authoritarian personalities, and Marcuse's theoretical work on instinct structure and society all follow the same conceptual strategy: basic psychological and sociological concepts can be interwoven because the perspectives projected in them of an autonomous ego and an emancipated society reciprocally require one another. This link of critical social theory to a concept of the ego that preserves the heritage of idealist philosophy in the no-longer idealist concepts of psychoanalysis is retained even when Adorno and Marcuse proclaim the obsolescence of psychoanalysis: "Society is beyond the stage at which psychoanalytic theory could illuminate its ingression into the psychic structure of the individual and could thereby reveal the mechanisms of social control *in* individuals. The cornerstone of psychoanalysis is the idea that social controls arise from the struggle between instinctual and social needs, from a struggle within the individual." [3] It is precisely this intrapsychic confrontation that is supposed to have become obsolete in the totally socialized society, which, so to speak, undercuts the family and directly imprints collective ego ideals on the child. Adorno had earlier argued in a similar vein: "Psychology is not a reservation for the particular protected from the general. The more social antagonisms increase, the more the thoroughly liberal and individualistic conception of psychology itself evidently loses its meaning. The pre-bourgeois world does not yet know psychology; the totally socialized world knows it no longer. To the latter corresponds analytic revisionism; this is adequate to the shift of power between society and the individual. Societal power hardly needs the mediating agencies of ego and individuality any longer. This then manifests itself as a growth of so-called ego psychology; while in truth individual psychological dynamics are replaced by the partly conscious, partly regressive adaptation of the individual to society." [4] But even this melancholy farewell to psychoanalysis appeals to the idea of an uncoerced ego that is identical with itself; how else could the form of total socialization be recognized, if not in the fact that it neither produces nor tolerates upright individuals.

I do not wish to go into the thesis of the end of the individual here.[5] In my view, Adorno and Marcuse have allowed themselves to be seduced, by an overly sensitive perception and an overly simplified interpretation of certain tendencies, into developing a left counterpart to the once-popular theory of totalitarian domination. I mention those utterances only to draw attention to the fact that critical social theory still holds fast to the concept of the autonomous ego, even when it makes the gloomy prognosis that this ego is losing its basis. Nonetheless, Adorno always refused to provide a direct explication of the normative content of basic critical concepts. To specify the make-up of the ego structures that are mutilated in the total society would have been regarded by him as false positivity. Adorno had good reasons to reject the demand for a positive conception of social emancipation and ego autonomy. He developed these reasons theoretically in his critique of First Philosophy: the attempts of ontological or anthropological thought to secure for themselves a normative foundation, as something first and unmediated, are doomed to failure. Additional reasons stem from the practical consideration that positive theories harbor a potential for legitimation that can be used, in opposition to their stated intentions, for purposes of exploitation and repression (as the example of classical doctrines of natural law shows). Finally, the normative content of basic critical concepts can be reconstructed nonontologically, that is, without recourse to a first unmediated something (or if you will, dialectically) only in the form of a developmental logic. But Adorno, despite his Hegelianism, distrusted the concept of a developmental logic because he held the openness and the initiative power of the historical process (of the species as well as of the individual) to be incompatible with the closed nature of an evolutionary pattern.

These are good reasons that can serve as a warning; but they can grant no dispensation from the duty of justifying concepts used with a critical intent. And Adorno did not always avoid doing so in philosophical contexts. In *Negative Dialectic* he says about the Kantian concept of the intelligible character: "According to the Kantian model subjects are free to the extent that they are conscious of themselves, are identical with themselves; and in such identity they are also unfree to the extent that they stand under and perpetuate its compulsion. As non-identical, as diffuse nature, they are unfree; and yet as such they are free, because in the impulses that overpower them they also become free of the compulsive character of identity." [6] I read this passage as an aporetic development of the determinations of an ego identity that makes freedom possible without demanding for it the price of unhappiness, violation of one's inner nature. I want to try to interpret this dialectical concept of ego identity with the cruder tools of sociological action theory and without fear of a false positivity; and I want to do so in such a way that the (no-longer-concealed) normative content can be incorporated in empirical theories and the proposed reconstruction of this content can be opened up to indirect testing.

II

The problems of development grouped around the concept of ego identity have been treated in three different theoretical traditions: in analytic ego psychology (H. S. Sullivan, Erikson), in cognitive developmental psychology (Piaget, Kohlberg), and in the symbolic interactionist theory of action (Mead, Blumer, Goffman, et al.).[7] If we step back for a moment and look for points of convergence among them, we find basic conceptions that can perhaps be summarized (in a simplified way) as follows.

1. The ability of the adult subject to speak and act is the result of the integration of maturational and learning processes, the interplay of which we cannot yet adequately understand. We can distinguish cognitive development from linguistic development and from psychosexual or motivational development. This motivational development seems to be intimately connected with the acquisition of interactive competence, that is, of the ability to take part in interactions (actions and discourses).[8]

2. The formative process of subjects capable of speaking and acting runs through an irreversible series of discrete and increasingly complex stages of development; no stage can be skipped over, and each higher stage implies the preceding stage in the sense of a rationally recon-

structible pattern of development. This concept of a developmental logic has been worked out especially by Piaget, but there are also certain correspondences in the other two theoretical traditions.[9]

3. The formative process is not only discontinuous but as a rule is crisis-ridden. The resolution of stage-specific developmental problems is preceded by a phase of destructuration and, in part, by regression. The experience of the productive resolution of a crisis, that is, of overcoming the dangers of pathological paths of development, is a condition for mastering later crises.[10] The concept of a maturational crisis has been worked out especially in psychoanalysis, but in connection with the adolescent phase it also has a meaning for the other two theoretical traditions.[11]

4. The developmental direction of the formative process is characterized by increasing autonomy. By that I mean the independence that the ego acquires through successful problemsolving, and through growing capabilities for problemsolving, in dealing with—

a) The reality of external nature and of a society that can be controlled from strategic points of view;

b) The nonobjectified symbolic structure of a partly internalized culture and society; and

c) The internal nature of culturally interpreted needs, of drives that are not amenable to communication, and of the body.[12]

5. The identity of the ego signifies the competence of a speaking and acting subject to satisfy certain consistency requirements. A provisional formulation by Erikson runs as follows: "The feeling of ego identity is the accumulated confidence that corresponding to the unity and continuity which one has in the eyes of others, there is an ability to sustain an inner unity and continuity." [13] Naturally ego identity is dependent on certain cognitive presuppositions; but it is not a determination of the epistemic ego. It consists rather in a competence that is formed in social interactions. Identity is produced through *socialization,* that is, through the fact that the growing child first of all integrates itself into a specific social system by appropriating symbolic generalities; it is later secured and developed through *individuation,* that is, precisely through a growing independence in relation to social systems.

6. The transposition of external structures into internal structures is an important learning mechanism. Piaget speaks of interiorization when schemata of action, that is, rules for the manipulative mastery of objects, are internally transposed and transformed into schemata of apprehension and of thought. Psychoanalysis and interactionism assert

a similar transposition of interaction patterns into intrapsychic patterns of relation (internalization).[14] With this mechanism is connected the further principle of achieving independence—whether from external objects, reference persons, or one's own impulses—by actively repeating what one has at first passively experienced or undergone.

In spite of these (admittedly somewhat stylized) convergent fundamental conceptions, none of these three theoretical approaches has as yet led to an explanatorily powerful theory of development, a theory that would permit a precise and empirically meaningful determination of the concept of ego identity (which is, nevertheless, being used more and more frequently). Taking analytic ego psychology as her point of departure, Jane Loevinger has, however, attempted to work out a theory that is meant to grasp ego development independently of cognitive development on the one side, and of psychosexual development on the other.[15] According to this conception, ego development and psychosexual development are together supposed to determine motivational development (see, Schema 1). I do not want to discuss this proposal in detail, but I shall point out three difficulties.

1. The claim to have grasped, in an analytically sharp way, something like ego development by employing the dimensions of behavioral control or superego formations, interactive style, and stage-specific developmental problems, does not strike me as plausible. For the developmental problems listed in the third column obviously do not lie in a single dimension, but touch on cognitive, motivational, and communicative tasks. Moreover, the superego formations circumscribed in the first column can scarcely be analyzed independently of psychosexual development.

2. The claim that the given stages of development follow an inner logic cannot be made good even intuitively. Nor does each row characterize a structural whole; nor can a hierarchy of increasingly complex stages of development building one on another be extracted from the columns.

3. Finally, the relation of the claimed logic of ego development to the empirical conditions under which it is realized in concrete life histories is not considered at all. Are there alternative paths of development that lead to the same goal? When do deviations occur from the rationally reconstructible developmental pattern? How great are the tolerance limits of the personality system and of social structures

Schema 1. *Stages of Ego Development*
(according to Jane Loevinger)

Stage	Impulse control and character development	Interpersonal style	Conscious preoccupation
Presocial Symbiotic		Autistic Symbiotic	Self vs. nonself
Impulse ridden	Impulse ridden, fear of retaliation	Exploitive; dependent	Bodily feelings, especially sexual and aggressive
Opportunistic	Expedient, fear of being caught	Exploitive, manipulative, zero-sum-game	Advantage, control
Conformist	Conformity to external rules, shame	Reciprocal, superficial	Things, appearance, reputation
Conscientious	Internalized rules, guilt	Intensive, responsible	Differentiated inner feelings, achievements, traits
Autonomous	Coping with inner conflict, toleration of differences	Intensive concern for autonomy	Ditto, role conceptualization, development, self-fulfillment
Integrated	Reconciling inner conflicts, renunciation of unattainable	Ditto, cherishing of individuality	Ditto, identity

Source: Jane Loevinger, "The Meaning and Measurement of Ego Development," *American Psychologist,* 21 (1966): 198.

Schema 1a. *Stages of Moral Consciousness*
(according to Lawrence Kohlberg)

Obedience and punishment orientation	Egocentric deference to superior power or prestige, or a trouble-avoiding set. Objective responsibility.	I Preconventional level
Instrumental hedonism	Right action is that instrumentally satisfying the self's needs and occasionally those of others. Naive egalitarianism and orientation to exchange and reciprocity.	
Good-boy orientation	Orientation to approval and to pleasing and helping others. Conformity to stereotypical images of majority or natural role behavior, and judgment by intentions.	II Conventional level
Law-and-order orientation	Orientation toward authority, fixed rules, and the maintenance of the social order. Right behavior consists of doing one's duty, showing respect for authority, and maintaining the given social order for its own sake.	
Contractual-legalistic orientation	Right action is defined in terms of individual rights and of standards which have been initially examined and agreed upon by the whole society. Concern with establishing and maintaining individual rights, equality, and liberty. Distinctions are made between values having universal, prescriptive applicability and values specific to a given society.	III Postconventional level
Universal-ethical-principle orientation	Right is defined by the decision of conscience in accord with self-chosen ethical principles appealing to logical comprehensiveness, universality, and consistency. These principles are abstract; they are not concrete moral rules. These are universal principles of justice, of the reciprocity and equality of human rights, and of respect for the dignity of human beings as individual persons.	

Source: Elliot Turiel, "Conflict and Transition in Adolescent Moral Development," *Child Development* 45 (1974): 14–29.

for such deviations? How do the stage of development and basic institutions of a society interfere with an ontogenetic developmental pattern?

I would like to deal with these difficulties in turn. First I shall isolate a central and well-examined aspect of ego development, namely moral consciousness. Even here I shall consider only the cognitive side, the ability to make moral judgments. (In Schema 1a I have coordinated the stages of moral consciousness proposed by Kohlberg with Schema 1, the stages of ego development proposed by Jane Loevinger, in order to emphasize that moral development represents a part of the development of personality that is decisive for ego identity.) I shall then show that Kohlberg's stages of moral consciousness satisfy the formal conditions for a developmental logic by reformulating these stages within a general action-theoretic framework. Last I shall remove the restriction to the cognitive side of communicative action and show that ego identity requires not only cognitive mastery of general levels of communication but also the ability to give one's own needs their due in these communication structures; as long as the ego is cut off from its internal nature and disavows the dependency on needs that still await suitable interpretations, freedom, no matter how much it is guided by principles, remains in truth unfree in relation to existing systems of norms.

III

Kohlberg defines six stages in a rationally reconstructible development of moral consciousness. To begin with, moral consciousness expresses itself in judgments about morally relevant conflicts of action. I call those action conflicts "morally relevant" that are capable of consensual resolution. The moral resolution of conflicts of action excludes the manifest employment of force as well as "cheap" compromises; it can be understood as a continuation of communicative action—that is, action oriented to reaching understanding—with discursive means. Thus the only resolutions permitted are those which:

> Harm the interests of at least one of the parties involved or affected;
> Nevertheless, permit a transitive ordering of the interests involved

from a point of view accepted as capable of consensus—the point of view, let us say, of a good and just life;

Entail sanctions in case of failure (punishment, shame, or guilt).

(Compare Kohlberg's definitions of the stages of moral consciousness in Schema 1b. As Schema 2 shows, different sanctions and domains of validity correspond to these stages.)

Schema 1b. *Definition of Moral Stages*

I. Preconventional level

At this level the child is responsive to cultural rules and labels of good and bad, right or wrong, but interprets these labels in terms of either the physical or the hedonistic consequences of action (punishment, reward, exchange of favors), or in terms of the physical power of those who enunciate the rules and labels. The level is divided into the following two stages:

Stage 1: *The punishment and obedience orientation.* The physical consequences of action determine its goodness or badness regardless of the human meaning or value of these consequences. Avoidance of punishment and unquestioning deference to power are valued in their own right, not in terms of respect for an underlying moral order supported by punishment and authority (the latter being stage 4).

Stage 2: *The instrumental relativist orientation.* Right action consists of that which instrumentally satisfies one's own needs and occasionally the needs of others. Human relations are viewed in terms like those of the market place. Elements of fairness, of reciprocity, and of equal sharing are present, but they are always interpreted in a physical pragmatic way. Reciprocity is a matter of »you scratch my back and I'll scratch yours«, not of loyalty, gratitude, or justice.

II. Conventional level

At this level, maintaining the expectations of the individual's family, group, or nation is perceived as valuable in its own right, regardless of immediate and obvious consequences. The attitude is not only one of *conformity* to personal expectations and social order, but of loyalty to it, of actively *maintaining,* supporting, and justifying the order, and of identifying with the persons or group involved in it. At this level, there are the following two stages:

Stage 3: *The interpersonal concordance or "good boy–nice girl" orientation.* Good behavior is that which pleases or helps others and is approved by them. There is much conformity to stereotypical images of what is majority or "natural" behavior. Behavior is frequently judged by intention—"he means well" becomes important for the first time. One earns approval by being "nice."

Stage 4: *The "law and order" orientation.* There is orientation toward authority, fixed rules, and the maintenance of the social order. Right behavior consists of doing one's duty, showing respect for authority, and maintaining the given social order for it's own sake.

III. Postconventional, autonomous, or principled level

At this level, there is a clear effort to define moral values and principles which have validity and application apart from the authority of the groups or persons holding these principles, and apart from the individual's own identification with these groups. This level again has two stages.

Stage 5: *The social-contract legalistic orientation,* generally with utilitarian overtones. Right action tends to be defined in terms of general individual rights, and standards which have been critically examined and agreed upon by the whole society. There is a clear awareness of the relativism of personal values and opinions and a corresponding emphasis upon procedural rules for reaching consensus. Aside from what is constitutionally and democratically agreed upon, the right is a matter of personal "values" and "opinion." The result is an emphasis upon the "legal point of view," but with an emphasis upon the possibility of changing law in terms of rational considerations of social utility (rather than freezing it in terms of stage 4 "law and order"). Outside the legal realm, free agreement and contract is the binding element of obligation. This is the "official" morality of the American government and constitution.

Stage 6: *The universal ethical principle orientation.* Right is defined by the decision of conscience in accord with self-chosen *ethical principles* appealing to logical comprehensiveness, universality, and consistency. These principles are abstract and ethical (the Golden Rule, the categorical imperative); they are not concrete moral rules like the Ten Commandments. At heart, these are universal principles of *justice,* of the *reciprocity* and *equality* of human *rights,* and of respect for the dignity of human beings as *individual persons.*

Source: Lawrence Kohlberg, "From Is to Ought," in T. Mishel, ed., *Cognitive Development and Epistemology* (New York, 1971), pp. 151–236.

Schema 2. Elucidation of the Stages of Moral Consciousness (Kohlberg)

Cognitive presuppositions	Stages of moral consciousness	Idea of the good and just life	Sanctions	Domain of validity
IIa. Concrete-operational thought	1. Punishment-obedience orientation	Maximization of pleasure through obedience	Punishment (deprivation of physical rewards)	Natural and social environment (undifferentiated)
	2. Instrumental hedonism	Maximization of pleasure through exchange of equivalents		
IIb. Concrete-operational thought	3. Good-boy orientation	Concrete morality of gratifying interactions	Shame (withdrawal of love and social recognition)	Group of primary reference persons
	4. Law-and-order orientation	Concrete morality of a customary system of norms		Members of the political community
III. Formal-operational thought	5. Social-contractual legalism	Civil liberty and public welfare	Guilt (reaction of conscience)	Legal associates in general
	6. Ethical-principled orientation	Moral freedom		Private persons in general

This empirically supported classification of expressions of moral judgment is supposed to satisfy the theoretical claim to represent developmental stages of moral consciousness. If we now take upon ourselves the burden of proof for this claim—a claim that Kohlberg himself has not made good—we commit ourselves to show that the descriptive sequence of moral types represents a developmental-logical nexus (in Flavell's sense). I should like to arrive at this goal through connecting moral consciousness with general qualifications for role behavior. The following three steps serve this end: first I introduce structures of possible communicative action and, indeed, in the sequence in which the child grows into this sector of the symbolic universe. I then coordinate with these basic structures the cognitive abilities (or competences) that the child must acquire in order to be able to move at the respective level of his social environment; that is, taking part first in incomplete interactions, then in complete interactions, and finally in communications that require passing from communicative action to discourse. Second, I want to look at this sequence of general qualifications for role behavior (at least provisionally) from developmental-logical points of view in order, finally, to derive the stages of moral consciousness from these stages of interactive competence.

I begin with the basic concepts of communicative action that must be presupposed for the perception of moral conflicts. These include. concrete behavioral expectations and corresponding intentional actions; then generalized behavioral expectations that are reciprocally connected with one another, that is, social roles and norms that regulate actions; principles that can serve to justify or to generate norms; the situational elements that are connected with actions (e.g., action consequences) or with norms (e.g., as conditions of application or as side effects); also actors who communicate with one another about something; and finally orientations, insofar as they are effective as motives for action. I am adopting the action-theoretic framework introduced by Mead and developed by Parsons, without thereby accepting conventional role theory.[16] (In Schema 3 I have ordered these components from the perspective of the socialization of the growing child.)

Schema 3.

| | General Structures of Communicative Action | | | | Qualifications of Role Behavior | | |
| | | | | | Perception of | | |
Cognitive presuppositions	Levels of interaction	Action levels	Action motivations	Actors	Norms	Motives	Actors
I Preoperational thought	Incomplete interaction	Concrete actions and consequences of action	Generalized pleasure/pain	Natural identity	Understand and follow behavioral expectations	Express and fulfill action intentions (wishes)	Perceive concrete actions and actors
II Concrete-operational thought	Complete interaction	Roles, systems of norms	Culturally interpreted needs	Role identity	Understand and follow reflexive behavioral expectations (norms)	Distinguish between "ought" and "want" (duty/inclination)	Distinguish between actions and norms, individual subjects and role bearers
III Formal-operational thought	Communicative action and discourse	Principles	Competing interpretations of needs	Ego identity	Understand and apply reflexive norms (principles)	Distinguish between heteronomy and autonomy	Distinguish between particular and general norms, individuality and ego in general

For the preschool child, who is cognitively still at the stage of preoperational thought, the sector of his symbolic universe relevant to action consists only of individual, concrete, behavioral expectations and actions, as well as consequences of action that can be understood as gratifications or sanctions. As soon as the child has learned to play social roles, that is, to participate in interactions as a competent member, his symbolic universe no longer consists only of actions that express concrete intentions, (e.g., wishes or wish fulfillments); rather, he can now understand actions as the fulfillment of temporally generalized behavioral expectations (or as violations of them). When, finally, the youth has learned to question the validity of social roles and norms of action, the sector of the symbolic universe expands once again; there now appear principles in accordance with which opposing norms can be judged. Dealing with hypothetical validity claims in this way requires the temporary suspension of constraints of action or, as we can also say, the entrance into discourses in which practical questions can be argumentatively clarified.

In the succession of these three levels, actors and their needs also grow stage-by-stage into the symbolic universe. At level I the orientations that guide action are integrated only to the extent that they can be generalized in the dimension of pleasure/pain. Only at level II is the satisfaction of need mediated through the symbolic devotion of primary reference persons, or through social recognition in expanded groups, in such a way that it is released from the egocentric tie to one's own balance of gratification. In this way, motives for action acquire the form of culturally interpreted needs; their satisfaction depends on following socially recognized expectations. At level III the quasi-natural process of need interpretation, which until then depended on an uncontrolled cultural tradition and changes in the institutional system, can itself be elevated to the object of discursive will-formation. In this way, beyond already culturally interpreted needs, the critique and justification of need interpretations acquire the power to orient action.

The stages through which the child grows into the general structures of communicative action have been described to a point at which there emerge corresponding indications for the perception and self-perception of actors, that is, of the subjects sustaining the interaction. When the child leaves its symbiotic phase and becomes sensitive to moral points of view—at first from the perspective of punishment and obedience—it has already learned to distinguish itself and its body from the environment, even though it does not yet strictly distinguish between physical and social objects in this environment. The child has thereby gained a "natural" identity, as it were, which it owes to the capacity of its body—as an organism that maintains boundaries—to conquer time. Plants and animals are already systems in an environment that possess not only an identity for us (the identifying observers), as do bodies-in-motion, but also an identity for themselves.[17] At the first level actors are thus not yet drawn into the symbolic world; there are natural agents to whom comprehensible intentions are ascribed, but not yet subjects whom one could *hold responsible* for actions with a view to generalized behavioral expectations. Only at the second level is identity detached from the bodily appearance of the actors. To the extent that the child assimilates the symbolic generalities of a few fundamental roles in his family environment, and later the norms of action of expanded groups, his natural identity is reformed through a symbolically supported role identity. Corporeal features such as sex, physical endowments, age, and so on, are absorbed into symbolic definitions. At this level actors appear as role-dependent reference persons and, later also, as anonymous role bearers. Only at the third level are the role bearers transformed into persons who can assert their identities independent of concrete roles and particular systems of norms. We are supposing here that the youth has acquired the important distinction between norms, on the one hand, and principles according to which we can generate norms, on the other—and thus the ability to judge according to principles. He takes into account that traditionally settled forms of life can prove to be mere conventions, to be irrational. Thus he has to retract his ego behind the line of all particular roles and norms and stabilize it only through the abstract ability to present himself credibly in any situation as someone who can satisfy the requirements of consistency even in the face of incompatible role expectations and in the passage through a sequence of contradictory periods of life. Role identity is replaced by ego identity; actors meet as individuals across, so to speak, the objective contexts of their lives.

Up to this point we have directed our attention to the components of the symbolic universe that acquire reality in stages for the growing child. If now, in a psychological attitude, we turn our attention to the abilities that the acting subjects must acquire in order to be able to move about in these structures, we come upon the general qualifications for role behavior that together form interactive competence. To the increasing mastery of the general structures of communicative action and the correlative growth of the acting subject's context-independence, there correspond graduated interactive competences that can be arranged in three dimensions (as shown on the right side of Schema 3). Our burden of proof will have been sufficiently discharged if the determinations introduced in each of these dimensions, regarded from a formal point of view, form a hierarchy such that the assertion of a developmental-logical nexus among the three levels of interaction can be justified.

The first dimension grasps the perception of the cognitive components of role qualifications: the actor must be able to understand and to follow the individual behavioral expectations of another (level I); he must be able to understand and to follow (or to deviate from) reflexive behavioral expectations— roles and norms (level II); finally he must be able to understand and apply reflexive norms (level III). The three levels are distinguished by degrees of reflexivity: the simple behavioral expectation of the first level becomes reflexive at the next level— expectations can be reciprocally expected; and the reflexive behavioral expectation of the second level again becomes reflexive at the third level—norms can be normed.

The second dimension relates to the perception of the motivational components of general role qualifications. At first the causality of nature is not distinguished from the causality of freedom—imperatives are understood in nature as well as in society as the expression of concrete wishes (level I); later the actor must be able to distinguish obligatory from merely desired actions (duty and inclination)—that is, the validity of a norm from the mere facticity of an expression of will (level II); and finally he must be able to distinguish between heteronomy and autonomy, that is, to see the difference between merely traditional (or imposed) norms and those which are justified in principle. The three levels are distinguished by degrees of abstraction and differentiation: the orientations that guide action become more and more abstract—from concrete needs through duties to the autonomous will—and at the same time more and more differentiated in regard to the validity claim of rightness (or "justice") that is connected with norms of action.

The third dimension grasps the perception of a component of general role qualifications which, if I am correct, presupposes the other two and has both cognitive and motivational sides. At first the actions and actors perceived are context-dependent, that is, concrete—there exists only the particular (level I). At the next level symbolic structures must be differentiated into general and particular—namely, individual actions vis-à-vis norms, and individual actors vis-à-vis role bearers. At the third level it must be possible to examine particular norms from the point of view of generalizability, so that the distinction between particular and general norms becomes possible. On the other side, actors can no longer be understood as a combination of role attributes; rather they count as individuated subjects who, through employing principles, can each organize an unmistakable biography. In other words, at this stage individuality and the "ego in general" [Ich überhaupt] must be differentiated. Here the levels are distinguished by degrees of generalization.

A glance at the columns I have just elucidated shows that role qualifications can be placed in a certain hierarchy from the formal viewpoints of (a) reflexivity, (b) abstraction and differentiation, and (c) generalization. This provides initial grounds for the conjecture that a deeper analysis could identify a developmental-logical pattern in Piaget's sense. In the present context, I shall have to let the matter rest with this conjecture. If it is correct, the same would have to hold for the stages of moral consciousness, insofar as these can be derived from the levels of role competence. This derivation as well can only be sketched here.

I shall proceed on the assumption that "moral consciousness" signifies the ability to make use of interactive competence for *consciously* processing morally relevant conflicts of action. You will recall that the consensual resolution of an action conflict requires a viewpoint that is open to consensus, with the aid of which a transitive ordering of the conflicting interests can be established. But competent agents will—independently of accidental commonalities of social origin, tradition, basic attitude, and so on—be in agreement about such a fundamental point of view only if it arises from the very structures of possible interaction. The reciprocity between acting subjects is such a point of view. In communicative action a relationship of at least incomplete reciprocity is established with the interpersonal relation between the involved parties. Two persons stand in an incompletely reciprocal relation insofar as one may do or expect x only to the extent that the other may do or expect y (e.g., teacher/pupil, parent/child). Their relationship is completely reciprocal if both may do or expect the same thing in comparable situations $(x = y)$ (e.g., the norms of civil law). In a now-famous essay Alvin Gouldner speaks of the norm of reciprocity that underlies all interactions.[18] This expression is not entirely apt, since reciprocity is not a norm but is fixed in the general structures of possible interaction. Thus the point of view of reciprocity belongs *eo ipso* to the interactive knowledge of speaking and acting subjects .

If this is granted, the stages of moral consciousness can be derived by applying the requirement of reciprocity to the action structures that the growing child perceives at each of the different levels (Schema 4). At level I, only concrete actions and action consequences (understood as gratifications or sanctions) can be morally relevant. If incomplete reciprocity is required here, we obtain Kohlberg's stage 1 (punishment-obedience orientation); complete reciprocity yields stage 2 (instrumental hedonism). At level II the sector relevant to action is expanded; if we require incomplete reciprocity for concrete expectations bound to reference persons, we obtain Kohlberg's stage 3 (good-boy orientation); the same requirement for systems of norms yields stage 4 (law-and-order orientation). At level III principles

Schema 4.

	Role Competence			Stages of Moral Consciousness				
Age level	Level of Communication		Reciprocity requirement	Stages of moral consciousness	Idea of the good life	Domain of validity	Philosophical reconstruction	Age level
I	Actions and consequences of action	Generalized pleasure/pain	Incomplete reciprocity	1	Maximization of pleasure—avoidance of pain through obedience	Natural and social environment		IIa
			Complete reciprocity	2	Maximization of pleasure—avoidance of pain through exchange of equivalents		Naive hedonism	
II	Roles	Culturally interpreted needs	Incomplete reciprocity	3	Concrete morality of primary groups	Group of primary reference persons		IIb
	Systems of norms	(Concrete duties)		4	Concrete morality of secondary groups	Members of the political community	Concrete thought in terms of a specific order	
III	Principles	Universalized pleasure/pain (utility)	Complete reciprocity	5	Civil liberties, public welfare	All legal associates	Rational natural law	III
		Universalized duties		6	Moral freedom	All humans as private persons	Formalistic ethics	
		Universalized need interpretations		7	Moral and political freedom	All as members of a fictive world society	Universal ethics of speech	

become the moral theme; for logical reasons complete reciprocity must be required. At this level the stages of moral consciousness are differentiated according to the degree to which action motives are symbolically structured. If the needs relevant to action are allowed to remain outside the symbolic universe, then the admissible universalistic norms of action have the character of rules for maximizing utility and general legal norms that give scope to the strategic pursuit of private interests, under the condition that the egoistic freedom of each is compatible with that of all. With this the egocentrism of the second stage is literally raised to a principle; this corresponds to Kohlberg's stage 5 (contractual-legalistic orientation). If needs are understood as culturally interpreted but ascribed to individuals as natural properties, the admissible universalistic norms of action have the character of general moral norms. Each individual is supposed to test monologically the generalizability of the norm in question. This corresponds to Kohlberg's stage 6 (conscience orientation). Only at the level of a universal ethics of speech [*Sprachethik*] can need interpretations themselves—that is, what each individual thinks he should understand and represent as his "true" interests—also become the object of practical discourse. Kohlberg does not differentiate this stage from stage 6, although there is a qualitative difference: the principle of justification of norms is no longer the monologically applicable principle of generalizability but the communally followed *procedure* of redeeming normative validity claims discursively. An unexpected result of our attempt to derive the stages of moral consciousness from the stages of interactive competence is the demonstration that Kohlberg's schema of stages is incomplete.

IV

A paradoxical relation is expressed in the identity of the ego: as a person in general the ego is like all other persons, but as an individual he is utterly different from all other individuals. Ego identity proves itself in the ability of the adult to construct new identities in conflict situations and to bring these into harmony

with older superseded identities so as to organize himself and
his interactions—under the guidance of general principles and
modes of procedure—into a unique life history. So far I have
developed only the cognitive and not the motivational side of
this concept of ego identity. I have chosen the perspective in
which we can observe how the ego of the child acquires in
stages the general structures of communicative action and, through
these, interactive competence, stability, and autonomy of action.
However this perspective screens out the psychodynamics of the
formative process. It neglects the instinctual processes into which
ego development is interwoven. In the dynamics of superego
formation, we can see the instrumental role that libidinous
energies, in the form of a narcissistic attachment to the self, play
in the development of ego ideals; we can also see the function
that aggressive energies, turned against the self, assume in the
establishment of the authority of conscience.[19] But above all, the
two major maturational crises—the Oedipal phase and adolescence—in which sex roles are learned and the motive-forming
powers of the cultural tradition are put to the test, show that the
ego can enter into and penetrate beyond structures of interaction
only if its needs can be admitted into and adequately interpreted
within the symbolic universe. In this perspective ego development presents itself as an extraordinarily dangerous process.
There is no need to refer to pathological developments to sub-
stantiate this fact; a less conspicuous sign, lying in the range of
the normal, are the frequent discrepancies between moral judgment and moral action.

The correlation between levels of interactive competence and
stages of moral consciousness (Schema 4) means that someone
who possesses interactive competence at a particular stage will
develop a moral consciousness at the same stage, insofar as his
motivational structure does not hinder him from maintaining,
even under stress, the structures of everyday action in the consensual regulation of action conflicts. In many cases, however,
the general qualifications for role behavior that are sufficient for
dealing with normal situations cannot be stabilized under the
stress of open conflicts. The party in question will then fall back
in his moral actions, or even in both his moral actions and moral

judgments, below the threshold of his interactive competence. There thus occurs a shifting between the stage of his normal role behavior and the stage at which he works through moral conflicts. Because it places the acting subject under an imperative for *consciously* working out conflicts, moral consciousness is an indicator of the degree of stability of general interactive competence.

The connection between conscious conflict resolution and morality becomes clear in extreme situations that do not admit an unequivocal moral solution, situations that make a rule violation (an offense) unavoidable. An action that nevertheless stands under conditions of morality in such situations is called "tragic." The concept of the tragic includes the intentional assumption of punishment or guilt, that is, the fulfillment of the moral postulate of consciousness even in the face of a morally insoluble dilemma. This throws some light on the meaning of moral action in general; we qualify as morally good those persons who maintain the interactive competence they have mastered for (largely conflict-free) normal situations even under stress, that is, in morally relevant conflicts of action, instead of unconsciously defending against conflict.

As ego psychology shows, the ego devises mechanisms for situations in which it would like to avoid conscious conflict resolution. These ingenious strategies for avoiding conflict contribute to a reaction to danger that is similar to flight; dangers are screened out of consciousness as the ego hides itself, as it were, from them. External reality and instinctual impulses are not the only sources of danger; the sanctions of the superego also represent a threat. We have anxiety if we act in moral conflicts otherwise than we believe by clear judgment that we have to act. In defending against these anxieties (which signal the recurrence of infantile anxieties) we conceal at the same time the discrepancy between our ability to judge and our willingness to act. The theory of defense mechanisms has, however, not been significantly improved since the first provisional attempt at systematization by Anna Freud.[20] Interestingly, several more recent investigations suggest that a developmental-logical ordering of the anxieties rekindled by transgression of moral commands (fear of punishment, shame, or qualms of conscience) makes possible a better classification of defense mechanisms.[21] Specific identity formations promote such anxieties because they make possible moral insights that are, so to speak, more advanced than the action motives that can be mobilized within their limits.

The dual status of ego identity reflects, of course, not only the cognitive-motivational duality of ego development but an interdependence of society and nature that extends into the formation of identity. The model of an unconstrained ego identity is richer and more ambitious than a model of autonomy developed exclusively from perspectives of morality. This can be seen in our completed hierarchy of the stages of moral consciousness. The meaning of the transition from the sixth to the seventh stage—in philosophical terms from a formalistic ethics of duty to a universal ethics of speech—can be found in the fact that need interpretations are no longer assumed as given, but are drawn into the discursive formation of will. Internal nature is thereby moved into a utopian perspective; that is, at this stage internal nature may no longer be merely examined within an interpretive framework fixed by the cultural tradition in a naturelike way, tested in the light of a monologically applied principle of generalization, and then split up into legitimate and illegitimate components, duties, and inclinations. Inner nature is rendered communicatively fluid and transparent to the extent that needs can, through aesthetic forms of expression, be kept articulable [sprachfähig] or be released from their paleosymbolic prelinguisticality. But that means that internal nature is not subjected, in the cultural preformation met with at any given time, to the demands of ego autonomy; rather, through a dependent ego it obtains free access to the interpretive possibilities of the cultural tradition. In the medium of value-forming and norm-forming communications into which aesthetic experiences enter, traditional cultural contents are no longer simply the stencils according to which needs are shaped; on the contrary, in this medium needs can seek and find adequate interpretations. Naturally this flow of communication requires sensitivity, breaking down barriers, dependency—in short, a cognitive style marked as field-dependent, which the ego, on the way to autonomy, first overcame and replaced with a field-*in*dependent style of perception and thought. Autonomy that robs the ego of a communicative access to its own inner nature also signals unfreedom. Ego identity means a freedom that limits itself in the intention of reconciling—if not of identifying—worthiness with happiness.

3   Historical Materialism and the Development of Normative Structures

This essay appeared as the introduction to *Zur Rekonstruktion des Historischen Materialismus*. Remarks referring to or based on the occasion have been omitted.

I

⌊In recent years I have made⌋ various attempts to develop a theoretical program that I understand as a reconstruction of historical materialism. The word *restoration* signifies the return to an initial situation that had meanwhile been corrupted; but my interest in Marx and Engels is not dogmatic, nor is it historical-philological. *Renaissance* signifies the renewal of a tradition that has been buried for some time; but Marxism is in no need of this. In the present connection, *reconstruction* signifies taking a theory apart and putting it back together again in a new form in order to attain more fully the goal it has set for itself. This is the normal way (in my opinion normal for Marxists too) of dealing with a theory that needs revision in many respects but whose potential for stimulation has still not been exhausted.

Not by chance [during the same period] I have been working on a theory of communicative action. Although the theory of communication is intended to solve problems that are rather of a philosophical nature—problems concerning the foundations of the social sciences—I see a close connection with questions relating to a theory of social evolution. This assertion might appear somewhat off the track; I would like, therefore, to begin by recalling the following circumstances:

a. In the theoretical tradition going back to Marx the danger of slipping into bad philosophy was always especially great when there was an inclination to suppress philosophical questions in favor of a scientistic understanding of science. Even in Marx himself the heritage of the philosophy of history sometimes came rather unreflectedly into play.[1] This historical objectivism took effect above all in the evolutionary theories of the Second International—for example, in Kautsky and "Diamat."[2] Thus special care is called for if we are today to take up once again the basic assumptions of historical materialism in regard to social evolution. This effort cannot consist in borrowing a list of prohibitions from a methodology developed with physics as the model, prohibitions that bar the way to social-scientific theories of development which pursue a research program suggested by Freud, Mead, Piaget, and Chomsky.[3] But care is called for in the choice of the basic concepts that determine the object domain of communicative action, for by this step, the kind of knowledge with which historical materialism may credit itself is decided.

b. From the beginning there was a lack of clarity concerning the normative foundation of Marxian social theory. This theory was not meant to renew the ontological claims of classical natural law, nor to vindicate the descriptive claims of nomological sciences; it was supposed to be "critical" social theory but only to the extent that it could avoid the naturalistic fallacies of implicitly evaluative theories. Marx believed he had solved this problem with a *coup de main,* namely, with a declaredly materialistic appropriation of the Hegelian logic.[4] Of course, he did not have to occupy himself especially with this task; for his practical research purposes he could be content to take at its word, and to criticize immanently, the normative content of the ruling bourgeois theories of modern natural law and political economy—a content that was, moreover, incorporated into the revolutionary bourgeois constitutions of the time. In the meantime, bourgeois consciousness has become cynical; as the social sciences—especially legal positivism, neoclassical economics, and recent political theory—show, it has been thoroughly emptied of binding normative contents. However, if (as becomes even more apparent in times of recession) the bourgeois ideals have gone into retirement, there are no norms and values to which an immanent critique might appeal with [the expectation of] agreement. On the other hand, the melodies of ethical socialism have been played through without result.[5] A philosophical ethics not restricted to metaethical statements is possible today only if we can reconstruct general presuppositions of communication and procedures for justifying norms and values.[6]

In practical discourse we thematize one of the validity claims that underlie speech as its *validity basis*. In action oriented to reaching understanding, validity claims are "always already" implicitly raised. These universal claims (to the comprehensibility of the symbolic expression, the truth of the propositional content, the truthfulness of the intentional expression, and the rightness of the speech act with respect to existing norms and values) are set in the general structures of possible communication. In these validity claims communication theory can locate a gentle but obstinate, a never silent although seldom redeemed claim to reason, a claim that must be recognized de facto whenever and wherever there is to be consensual action.[7] If this is idealism, then idealism belongs in a most natural way to the conditions of reproduction of a species that must preserve its life through labor and interaction, that is, *also* by virtue of propositions that can be true and norms that are in need of justification.[8]

c. Not only are there connections between the theory of communicative action and the foundations of historical materialism; in examining individual assumptions of evolutionary theory, we run up against problems that make communications-theoretical reflections necessary. Whereas Marx localized the learning processes important for evolution in the dimension of objectivating thought—of technical and organizational knowledge, of instrumental and strategic action, in short, of *productive forces*—there are good reasons meanwhile for assuming that learning processes also take place in the dimension of moral insight, practical knowledge, communicative action, and the consensual regulation of action conflicts—learning processes that are deposited in more mature forms of social integration, in new *productive relations,* and that in turn first make possible the introduction of new productive forces. The rationality structures that find expression in world views, moral representations, and identity formations, that become practically effective in social movements and are finally embodied in institutional systems, thereby gain a strategically important position from a theoretical point of view. The systematically reconstructible patterns of development of normative structures are now of particular interest. These structural patterns depict a *developmental logic* inherent in cultural traditions and institutional change. This logic says nothing about the *mechanisms* of development; it says something only about the range of variations within which cultural values, moral representations, norms, and the like—at a given level of social organization—can be changed and can find different historical expression. In its developmental *dynamics,* the change of normative structures remains dependent on evolutionary challenges posed by unresolved, economically conditioned, system problems and on learning processes that are a response to them. In other words, culture remains a superstructural phenomenon, even if it does seem to play a more prominent role in the transition to new developmental levels than many Marxists have heretofore supposed. This prominence explains the contribution that communication theory can, in my view, make to a renewed historical materialism. In the following pages I would like to at least suggest wherein this contribution could consist.

II

The structures of linguistically established intersubjectivity— which can be examined prototypically in connection with elementary speech actions—are conditions of both social and personality systems. Social systems can be viewed as networks of communicative actions; personality systems can be regarded under the aspect of the ability to speak and act. If one examines social institutions and the action competences of socialized individuals for general characteristics, one encounters the same structures of consciousness. This can be shown in connection with those arrangements and orientations that specialize in maintaining the endangered intersubjectivity of understanding in cases of action conflicts—law and morality. When the background consensus of habitual daily routine breaks down, consensual regulation of action conflicts (accomplished under the renunciation of force) provides for the continuation of communicative action with other means. To this extent, law and morality mark the core domain of interaction. One can see here the identity of the conscious structures that are, on the one hand, embodied in the institutions of law and morality and that are, on the other hand, expressed in the moral judgments and actions of individuals. Cognitive developmental psychology has shown that in ontogenesis there are different stages of moral consciousness, stages that can be described in particular as preconventional, and postconventional patterns of problemsolving.[9] The same patterns turn up again in the social evolution of moral and legal representations.

The ontogenetic models are certainly better analyzed and better corroborated than their social-evolutionary counterparts. But it should not surprise us that there are homologous structures of consciousness in the history of the species, if we consider that linguistically established intersubjectivity of understanding marks that innovation in the history of the species which first made possible the level of sociocultural learning. At this level the reproduction of society and the socialization of its members are two aspects of the same process; they are dependent on the same structures.

The homologous structures of consciousness in the histories of the individual and the species [are not restricted to the domain of law and morality]. The success of the theoretical approach programmatically presented here also requires an investigation of rationality structures in domains that have heretofore been scarcely examined, either conceptually or empirically—in the domain of ego development and the evolution of world views on the one hand, and in the domain of ego and group identities on the other.[10]

To begin with, the *concept of ego development,* ontogenesis, can be analyzed in terms of the capability for cognition, speech, and action. These three aspects of cognitive, linguistic, and interactive development can be brought under one unifying idea of ego development—the ego is formed in a system of demarcations. The subjectivity of internal nature demarcates itself in relation to the objectivity of a perceptible external nature, in relation to the normativity of society, and in relation to the intersubjectivity of language. In accomplishing these demarcations, the ego knows itself not only as subjectivity but as something that has "always already" transcended the bounds of subjectivity in cognition, speech, and interaction simultaneously. The ego can identify with itself precisely in distinguishing the merely subjective from the nonsubjective. From Hegel through Freud to Piaget, the idea has developed that subject and object are reciprocally constituted, that the subject can grasp hold of itself only in relation to and by way of the construction of an objective world. This nonsubjective is, on the one hand, an "object" in Piaget's sense—a cognitively objectified and manipulable reality; on the other hand, it is an "object" in Freud's sense—a domain of interaction opened up by communication and secured through identification. The environment is differentiated into these two regions: external nature and society. It is supplemented by reflections of the two domains of reality in each other (e.g., nature as "fraternal," cared for on an analogy to society; or society as a strategic game, as a system, and so forth). In addition, language detaches itself from the domains of objects as a region unto itself.

Psychoanalytic and cognitive developmental psychology have assembled evidence for the hypothesis that ego development takes place in stages. I should like—very tentatively—to distinguish among (a) the symbiotic, (b) the egocentric, (c) the sociocentric-objectivistic, and (d) the universalistic stages of development.[11]

a. During the first year of life we can find no clear indicators for a subjective separation between subject and object. Apparently in this phase the child cannot perceive its own corporeal substance as a body, as a boundary-maintaining system. The symbiosis between child, reference person, and physical environment is so intimate that we cannot meaningfully speak of a demarcation of subjectivity in the strict sense.

b. In the next segment of life, which corresponds with Piaget's sensory-motor and preoperative phases of development, the child succeeds in differentiating between self and environment. It learns to perceive permanent objects in its environment, but without yet clearly differentiating the environment into physical and social domains. Moreover, the demarcation [of the self] in relation to the environment is not yet objective. This can be seen in manifestations of cognitive and moral egocentrism. The child cannot perceive, understand, and judge situations independently of its own standpoint—it thinks and acts from a body-bound perspective.

c. With the onset of the stage of concrete operations the child takes the decisive step toward constructing a system of demarcations; it now differentiates between perceptible and manipulable things and events, on the one hand, and understandable action-subjects and their utterances, on the other; and it no longer confuses linguistic signs with the reference and meaning of symbols. In becoming aware of the perspectival character of its own standpoint, it learns to demarcate its subjectivity in relation to external nature and society. With the seventh year, more or less, pseudo-lying ceases—an indication that distinctions are made between fantasies and perceptions, impulses and obligations. At the close of this phase, cognitive development has led to an objectivation of external nature, linguistic-communicative development to the mastery of a system of speech acts, and interactive development to the complementary connection of generalized expectations of behavior.

d. Only with adolescence can the youth succeed in progressively freeing himself from the dogmatism of the preceding phase of development. With the ability to think hypothetically and to conduct discourses, the system of ego-demarcations becomes reflective. Until then the epistemic ego, bound to concrete operations, confronted an objectivated nature; and the practical ego, immersed in group perspectives, was dissolved in quasi-natural systems of norms. But when the youth no longer naively accepts the validity claims contained in assertions and norms, he can transcend the objectivism of a given nature and, in the light of hypotheses, explain the given from contingent boundary conditions; and he can burst the sociocentrism of a traditional order and, in the light of principles, understand (and if necessary criticize) existing norms as mere conventions. To the extent that the dogmatism of the *given* and the *existing* is broken, the pre-scientifically constituted object domains can be relativized in relation to the system of ego-demarcations so that theories can be traced back to the cognitive accomplishments of investigating subjects and norm systems to the will-formation of subjects living together.

If we go on now to seek homologies between ego development and the evolution of world-views, we must take care not to draw hasty parallels.

a. The confusion of structure and content can easily lead to errors—individual consciousness and cultural tradition can agree in their content without expressing the same structures of consciousness.

b. Not all individuals are equally representative of the developmental stage of their society. Thus in modern societies, law has a universalistic structure, although many members are not in a position to judge according to principles. Conversely, in archaic societies there were individuals who had mastered formal operations of thought, although the collectively shared mythological world-view corresponded to a lower stage of cognitive development.

c. The ontogenetic pattern of development cannot mirror the structures of species history for the obvious reason that collective structures of consciousness hold only for adult members—ontogenetically early stages of incomplete interaction have no correspondents, even in the oldest societies, for (with the family organization) social relations have had from the beginning the form of complementarily connected, generalized expectations of behavior (i.e., the form of complete interaction).

d. Furthermore, the points of reference from which the same structures of consciousness are embodied are different in the history of the individual and in that of the species. The maintenance of the personality system poses quite different imperatives than the maintenance of the social system.

e. There is a special proviso for structural comparison of ego development and world-view development. The unifying power of world-views is directed not only against cognitive dissonance but also against social disintegration. The concordant structuration of the stock of knowledge stored and harmonized in interpretive systems is related, therefore, not only to the unity of the epistemic ego, but also to that of the practical ego. Legal and moral representations are to be distinguished in turn from the concepts and structures that directly serve to stabilize ego and group identities—for example, concepts of originary powers, gods, representations of the soul, concepts of fate, and the like. This complex construction prohibits a global comparison between ego development and the development of world views. We have to sharpen particular, abstract, reference points of comparison. Thus there might be a process of decentration of world views that corresponds to ego development; and for cognitive development in the narrower sense, we can look for isomorphisms in the fundamental concepts and logical connections of collective interpretive systems.

All provisos notwithstanding, certain homologies can be found. This is true in the first place for cognitive development. In ontogenesis we can observe sequences of basic concepts and logical structures similar to those observable in the evolution of world views[12]—for example, the differentiation of temporal horizons and the separation of physically measured and biographically experienced time; the articulation of a concept of causality— grasped only globally at first—that becomes specified for the causal connection of things and events, on the one side, and for the motivational connection of actions, on the other, and is later employed as a basis for the hypothetical concepts of a law of nature and a norm of action; or the differentiation of the concept of substance—encompassing at first the animate and the inanimate—into objects that can be manipulated and social objects that can be encountered as opposite numbers in interaction. (Thus, for example, Döbert has attempted to reconstruct the development of religion from primitive mythology to so-called modern religion—which has shrunk to a profane ethics of communication—from the point of view of a step-by-step explication of basic action-theoretic concepts.)[13] Similar observations can be made regarding logical structures. Mythology permits narrative explanations with the help of exemplary stories; cosmological world views, philosophies, and higher religions already permit deductive explanations from first principles (the originary actions of myth having been transformed into "beginnings" of argumentation, beyond which one cannot go); modern science, finally, permits nomological explanations and practical justifications, with the help of revisable theories and constructions that are monitored against experience. When these various types of explanation (and justification) are analyzed formally, we find developmental-logical correlations with ontogenesis. In the present connection, however, we are less interested in the structural analogies between world views and cognitive (in the narrower sense) development than in those between world views and *the system of ego demarcations*.

Apparently the magical-animistic representational world of paleolithic societies was very particularistic and not very coherent. The ordering representations of mythology first made possible the construction of a complex of analogies in which all natural and social phenomena were interwoven and could be transformed into one another. In the egocentric world conception of the child at the preoperational level of thought, these phenomena are made relative to the center of the child's ego; similarly, in sociomorphic world views they are made relative to the center of the tribal group. This does not mean that the members of the group have formed a distinct consciousness of the normative reality of a society standing apart from objectivated nature—these two regions have not yet been clearly separated. Only with the transition to societies organized around a state do mythological world views also take on the legitimation of structures of domination (which already presuppose the conventional stage of moralized law). Thus the naive attitude to myth must have changed by that time. Within a more strongly differentiated temporal horizon, myth is distantiated to a tradition that stands out from the normative reality of society and from a partially objectivated nature. With persisting sociomorphic traits, these developed myths establish a unity in the manifold of appearances; in formal respects, this unity resembles the sociocentric-objectivistic world conception of the child at the stage of concrete operations.

The further transition from archaic to developed civilizations is marked by a break with mythological thought. There arise cosmological world views, philosophies, and the higher religions, which replace the narrative explanations of mythological accounts with argumentative foundations. The traditions going back to the great founders are an explicitly teachable knowledge that can be dogmatized, that is, professionally rationalized. In their articulated forms rationalized world views are an expression of formal-operational thought and of a moral consciousness guided by principles. The cosmologically or monotheistically conceived totality of the world corresponds formally to the unity that the youth can establish at the stage of universalism. Of course, the universalistic structures of world views have to be made compatible with the traditionalistic attitude toward the political order that predominates in the ancient empires. This is possible above all because the highest principles, to which all argumentation recurs, are themselves removed from argumentation and immunized against objections. In the ontological tradition of thought, this finality is guaranteed through the concept of the absolute (or of complete self-sufficiency).

In the course of the establishment of universalistic forms of intercourse in the capitalist economy and in the modern state the attitude toward the Judaeo-Christian and Greek-ontological heritage was refracted in a subjectivistic direction (the Reformation and modern philosophy). The highest principles lost their unquestionable character; religious faith and the theoretical attitude became reflective. The advance of the modern sciences and the development of moral-practical will-formation were no longer prejudiced by an order that—although grounded—was posited absolutely. For the first time, the universalistic potential already contained in the rationalized world views could be set free. The unity of the world could no longer be secured objectively, through hypostasizing unifying principles (God, Being, or Nature); henceforth it could be asserted only reflectively, through the unity of reason (or through a rational organization of the world, the actualization of reason). The unity of theoretical and practical reason then became the key problem for modern world interpretations, which have lost their character *as* world views.

These fleeting allusions are meant only to render plausible the heuristic fruitfulness of the conjecture that there are homologies between the structures of the ego and of world-views. In both dimensions, development apparently leads to a growing decentration of interpretive systems[14] and to an ever-clearer categorical demarcation of the subjectivity of internal nature from the objectivity of external nature, as well as from the normativity of social reality and the intersubjectivity of linguistic reality.

III

There are also homologies between the structures of ego identity and of group identity. The epistemic ego (as the ego in general) is characterized by those general structures of cognitive, linguistic, and active ability that every individual ego has in common with all other egos; the practical ego, however, forms and maintains itself as individual in performing its actions. It secures the identity of the person within the epistemic structures of the ego in general. It maintains the continuity of life history and the symbolic boundaries of the personality system through repeatedly actualized self-identifications; and it does so in such a way that it can locate itself clearly—that is, unmistakably and recognizedly—in the intersubjective relations of its social life world. Indeed the identity of the person is in a certain way the result of identifying achievements of the person himself.[15]

In our propositional attitude toward things and events (and derivatively also toward persons and their utterance)—that is, in making (or understanding) a statement about them—we undertake an identification. For that purpose we employ names, characterizations, demonstrative pronouns, and so on. Deictic expressions (and gestures) contain identifying features that suffice in a given context to single out a particular—indeed the intended—object from a class of similar objects (e.g., to distinguish *this* stone, about which I want to assert something, from all other stones). Spatio-temporal positions are the most abstract features suitable for identifying any bodies whatever. Persons too can be identified in a propositional attitude, for example, in connection with corporeal features such as size, hair and eye color, scars, fingerprints, and so on.[16] But in difficult cases these criminological characteristics are not sufficient; in extreme cases we are left with requesting the person in question to clarify his own identity. As long as he denies the identity propositionally ascribed to him, we cannot be certain whether he is simply disavowing his identity or is not in a position to sustain his identity (whether he has perhaps a split personality), or whether he is not at all the person we suspect he is on the basis of external characteristics. There may be world-shaking evidence for the *bodily identity* of a person; but to attain certainty regarding the *identity of the person*, we must give up our propositional attitude and, in a performative attitude, ask the one involved about his identity, ask him to identify *himself*. In doubtful cases we have to identify other persons according to the characteristics through which they identify themselves.

No one can construct an identity independently of the identifications that others make of him.[17] These are, naturally, identifications that others make not in the propositional attitude of observers, but in the performative attitude of participants in interaction. Indeed the ego does not accomplish its self-identifications in a propositional attitude. It presents itself to itself as a practical ego in the performance of communicative actions; and in communicative action the participants must reciprocally suppose that the distinguishing-oneself-from-others is recognized by those others. Thus the basis for the assertion of one's own identity is not really self-identification, but intersubjectively recognized self-identification.

The expressions *I* and *thou* do not—as the personal pronouns in the third person do—have the referential meaning of denotative expressions employed propositionally. Rather they borrow their referential meaning from the illocutionary roles of linguistic performance; they have primarily the meaning of personal self-representation on the basis of the intersubjective recognition of reciprocal self-representations. The expressions *we* and *you* have the same relation to personal pronouns of the third-person plural as do *I* and *thou* to those of the third-person singular. But there is nonetheless an interesting asymmetry. The expression "we" is used not only in collective speech actions vis-à-vis an addressee who assumes the communicative role of *you*, under the reciprocity condition that *we* in turn are *you* for them. In individual speech actions, *we* can also be used in such a way that a corresponding sentence presupposes not the complementary relation to another group but that to other individuals of one's own group.

    1) We took part in the demonstration (while you sat home).

    2) We are all in the same boat.

Sentence (1) is addressed to another group, sentence (2) to members of one's own group. Sentences of type (2) have not only the usual self-referential meaning, but the meaning of self-identification—we are X (where X can signify Germans, citizens of Hamburg, women, redheads, workers, blacks, and so on). The expression *I* can also be used for purposes of self-identification; but the self-identification of an *I* requires intersubjective recognition by other *I*'s, who must in turn assume the role of *thou*. By contrast, the self-identification of a group is not dependent on intersubjective recognition by *another* group; an *I* that identifies itself as *we* can be confirmed through another *I* that identifies with the same *we*. The reciprocal recognition of group members requires I-thou-we relations.

This has consequences for the construction of a collective identity. I would like to reserve the expression *collective identity* for reference groups that are essential to the identity of their members, which are in a certain way "ascribed" to individuals, cannot be freely chosen by them, and which have a continuity that extends beyond the life-historical perspectives of their members. For the construction of such identities, I-thou-we relations are sufficient; we-you relations are not a necessary condition (as I-thou relations are for the construction of a personal identity). In other words, a group can understand and define itself so exclusively as a totality that they live in the idea of embracing all possible participants in interaction, whereas everything that doesn't belong thereto becomes a neuter, about which one can make statements in the third person, but with which one cannot take up interpersonal relations in the strict sense—as was the case, for instance, with the barbarians on the borders of the ancient civilizations. I cannot here go any further into the logic of the use of personal pronouns—a logic that is the key to the concept of identity[18]—but I do want to call briefly to mind the ontogenetic stages of identity formation, in order to render precise the sense in which ego identity is understood as the ability to sustain one's own identity.

I distinguished between the identity that is propositionally ascribed to things and events and the identity that persons claim for themselves and maintain in communicative action. I did not mention the identity of boundary-maintaining organisms, which have an identity not only "for us," as observers, but also an identity "for themselves," without, however, being able to represent and to secure it in the medium of linguistically established intersubjectivity. (In his important book on the *Stufen des Organischen* (1928), Helmuth Plessner—employing a conceptual apparatus influenced by Fichte's philosophy of reflection—tried to distinguish different "positionalities," and to clarify the concept of the natural identity of living beings.) The "natural identity" of early childhood is probably also based on the time-conquering character of a boundary-maintaining organism, namely, the child's own body, which it gradually learns to distinguish from the physical/social environment. By contrast, the unity of the person, which is constructed by way of intersubjectively recognized self-identification (analyzed by G. H. Mead), is based on belonging to, and demarcating oneself from, the symbolic reality of a group, and on the possibility of locating oneself in it. The unity of the person is formed through internalization of roles that are originally attached to concrete reference persons and later detached from them—primarily the generation and sex roles that determine the structure of the family. This role identity, centered on sex and age and integrated with the child's own body image, becomes more abstract and, at the same time, more individual to the degree that the young child appropriates extra-familial role systems up to and including the political order, which is interpreted and justified by a complex tradition.

The continuity-guaranteeing character of role identities is based on the intersubjective validity and temporal stability of behavioral expectations. If the development of moral consciousness leads beyond this conventional stage, role identity is shattered because the ego then withdraws behind all particular roles. An ego expected to judge any given norm in the light of internalized principles, that is, to consider them hypothetically and to provide justifications, can no longer tie its identity to particular pregiven roles and sets of norms.[19] Now continuity can be established only through the ego's own integrating accomplishment. This ability is paradigmatically exercised when the growing child gives up its early identities, which are tied to familial roles, in favor of more and more abstract identities secured finally to the institutions and traditions of the political community. To the extent that the ego generalizes this ability to overcome an old identity and to construct a new one and learns to resolve identity crises by reestablishing at a higher level the disturbed balance between itself and a changed social reality, role identity is replaced by ego identity. The ego can then maintain his identity in relation to others, expressing in all relevant role games the paradoxical relationship of being like and yet being absolutely different from the other, and represent himself as the one who organizes his interactions in an unmistakable complex of life history.[20]

In modern society this ego identity could be supported by individualistic vocational roles. The vocational role, understood in Weber's sense, was the most significant vehicle for projecting a unifying life-historical career. Today this vehicle seems more and more to be slipping away. Thus feminism is an example of an emancipatory movement that (under the catchword of self-realization) searches for paradigmatic solutions to the problem of stabilizing ego identity under conditions that render problematic—especially for women—recourse to the vocational role as the crystallizing nucleus of life history.

In looking for homologies between patterns of identity development and the historical articulation of collective identities, we have to avoid, again, drawing hasty parallels. The provisos I mentioned above hold here as well; and I would like to add three special ones.

a. The collective identity of a group or a society secures continuity and recognizability. For this reason it varies with the temporal concepts in terms of which the society can specify the requirements for remaining the same. The individual lifetime too is schematized on different levels of cognitive development; but it is objectively bounded, at least by birth and death. There are no comparable objective cut-off points for the historical existence of a society, with its overreaching generations and, often, epochs.

b. Furthermore, collective identity determines how a society demarcates itself from its natural and social environments. In this respect too, clear analogies [to individual life] are lacking. A personal life-world is bounded by the horizon of all possible experiences and actions that can be attributed to the individual in his exchange with his social environment. By contrast, the symbolic boundaries of a society are formed primarily as the horizon of the actions that members reciprocally attribute to themselves internally.

c. The third feature is all the more important—collective identity regulates the membership of individuals in the society (and exclusion therefrom). In this respect there is a complementary relation between ego and group identity, because the unity of the person is formed through relations to other persons of the same group; and as I mentioned above, identity development is characterized by the fact that early identification with concrete and less complex groups (the family) is weakened and subordinated to identification with more encompassing and more abstract units (city, state). This suggests that we can infer from the ontogenetic stages of ego development the complementary social structures of the tribal group, the state, and, finally, global forms of intercourse. Elsewhere I advanced certain conjectures to that effect; [21] but I see now that I underestimated the complexity of the connection of collective identities with world views and systems of norms. Following Parsons, we can distinguish cultural values, actions systems in which values are institutionalized, and collectives that act in these systems. Only *a certain segment* of the culture and action system is important for the identity of a collective—namely the taken-for-granted, consensual, basic values and institutions that enjoy a kind of fundamental validity in the group. Individual members of the group perforce experience the destruction or violation of this normative core as a threat to their own identity. The different forms of collective identity can be found only in such normative cores, in which individual members "know themselves as one" with each other.

In *neolithic societies* collective identity was secured through the fact that individuals traced their descent to the figure of a common ancestor and thus, in the framework of their mythological world-view, assured themselves of a common cosmogonic origin. On the other hand, the personal identity of the individual developed through identification with a tribal group, which was in turn perceived as part of a nature interpreted in interaction categories. As social reality was not yet clearly distinguished from natural reality, the boundaries of the social world merged into those of the world in general.[22] Without clearly defined boundaries of the social system there was no natural or social environment in the strict sense; contacts with alien tribes were interpreted in accord with the familiar kinship connection. On the other hand, encounters with civilizations that (unlike alien tribes) could no longer be assimilated to their own world represented a danger for the collective identity of societies organized along kinship lines (independently of the real danger of colonial conquest).[23]

The transition to *societies organized through a state* required the relativization of tribal identities and the construction of a more abstract identity that no longer based the membership of individuals on common descent but on belonging in common to a territorial organization. This took place first through identification with the figure of a ruler who could claim a close connection and privileged access to the mythological originary powers. In the framework of mythological world views the integration of different tribal traditions was accomplished through a large-scale, syncretistic expansion of the world of the gods—a solution that proved to be rather unstable. For this reason, imperially developed civilizations had to secure their collective identity in a way that presupposed a break with mythological thought. The universalistic world interpretations of the great founders of religions and of the great philosophers grounded a commonality of conviction mediated through a teaching tradition and permitting only abstract objects of identification. As members of universal communities of faith, citizens could recognize their ruler and the order represented by him so long as it was possible to render political domination plausible in some sense as the legacy of an order of the world and of salvation that was believed in and posited absolutely.

In contrast to archaic tribal societies, the *great empires* had to demarcate themselves from a desocialized outer nature as well as from the social environment of those alien to the empire. But since collective identity could now be secured only by way of doctrines with a universal claim, the political order also had to be in accord with this claim—the empires were not universal in name alone. Their peripheries were fluid; they consisted of allies and dependents. In addition there were barbarians, whom one attempted either to conquer or to convert—aliens who were potential members but who, so long as they had not the status of citizen, did not count as fully human. Only the reality of other empires was incompatible with this definition of the boundaries and social environment of an empire. Despite the existence of trade relations, and despite the diffusion of innovations, the empires shielded themselves from this danger; among themselves they maintained no diplomatic relations in the sense of an institutionalized foreign policy. In any case, their political existence was not dependent on a system of reciprocal recognition.

The limits of this identity formation manifested themselves inwardly as well. In societies organized along kinship lines collective identity was correlated, in most cases, with individual role identities established through kinship structures. Within the framework of mythological world views there was no stimulus to develop identity beyond this stage; individual discrepancies could easily be accommodated in the roles of priest and shaman.[24] In highly stratified civilizations, on the other hand, the integrating power of the identity of the empire had to confirm itself precisely in unifying the evolutionarily nonsynchronous structures of consciousness of the country, the aristocracy, city tradesmen, priests, and officials, and in binding them to the same political order. A broad spectrum of belief attitudes toward the same tradition was permitted; what was for one something like a myth that could still be connected with magical practices was for others a tradition of faith, however supported by ritual. The dogmatic organization of doctrinal knowledge often displaced even the weight of tradition with the weight of arguments and replaced an attitude of faith based on the authority of a doctrine with a theoretical attitude. But this universalistic potential could not be released on a large scale if the particularity of domination and of the citizen's status, which was merely concealed by the empire's claim to universality, was to remain imperceptible and not lead to significant discrepancies.

Such discrepancies turned up again and again in the ancient empires; but only with the transition to *the modern world* did they become unavoidable. The capitalist principle of organization meant the differentiation of a depoliticized and market-regulated economic system. This domain of decentralized individual decisions was organized on universalistic principles in the framework of bourgeois civil law. It was thereby supposed that the private, autonomous, legal subjects pursued their interests in this morally neutralized domain of intercourse in a purposive-rational manner, in accord with general maxims.[25] From this conversion of the productive sphere to universalistic orientations there proceeded a strong structural compulsion for the development of personality structures that replaced conventional role identity with ego identity. In fact, emancipated members of bourgeois society, whose conventional identity had been shattered, could know themselves as one with their fellow citizens in their character as (a) free and equal subjects of civil law (the citizen as private commodity owner), (b) morally free subjects (the citizen as private person), and (c) politically free subjects (the citizen as democratic citizen of the state).[26] Thus the collective identity of bourgeois society developed under the highly abstract viewpoints of legality, morality, and sovereignty; at least it expressed itself in this way in modern natural-law constructions and in formalist ethics.

However, these abstract determinations are best suited to the identity of world citizens, not to that of citizens of a particular state that has to maintain itself against other states. The modern state arose in the sixteenth century as a member of a system of states; the sovereignty of one found its limits in the sovereignty of all other states; indeed its sovereignty was only constituted in this system based on reciprocal recognition. Even if the system could have defined away, as peripheral, the non-European world with which it was economically involved from the start, it still could not have represented itself as a universal unity in the style of a grand empire. This was excluded by the international relations between the sovereign states—relations based in the final analysis on the threat of military force. Moreover, the modern state was more reliant than the state in tribal societies on the loyalty and willingness to sacrifice of a population made economically and socially mobile. The identity of world citizens obviously is not strong enough to establish universal conscription. A symptom of this can be seen in the double identity of the citizen of the modern state—he is *homme* and *citoyen* in one.[27]

This competition between two group identities was temporarily silenced through membership in nations: the nation is the modern identity formation that defused and made bearable the contradiction between the intrastate universalism of bourgeois law and morality, on the one side, and the particularism of individual states, on the other. Today there are a number of indications that this historically significant solution is no longer stable. The Federal Republic of Germany has the first army expected by the responsible minister to maintain fighting readiness without an image of the enemy.[28] Conflicts that are ignited below the threshold of national identity are breaking out everywhere, in connection with questions of race, creed, language, regional differences, and other subcultures.[29]

One alternative to the presently disintegrating national identity was the European working-class movement. Taking the bourgeois philosophy of history as its point of departure, historical materialism projected a collective identity compatible with universalistic ego structures. What the eighteenth century had thought of under the rubric of cosmopolitanism was now conceived of as socialism; but this identity was projected into the future and thus made a task for political practice. This was the first example of an identity that had become reflective, of a collective identity no longer tied retrospectively to specific doctrines and forms of life but prospectively to programs and rules for bringing about something. Until now identity formation of this type could be maintained only in social movements; whether societies in a normal state could develop such a fluid identity is questionable. It would have to adjust itself to high mobility, not only in regard to productive resources, but also in regard to processes of norm and value formation. For the time being only China is experimenting with such arrangements.

This sketch can at best suggest how to use the identity development of the individual as a key to the change of collective identities. In both dimensions identity projections apparently become more and more general and abstract, until finally the projection mechanism as such becomes conscious, and identity formation takes on a reflective form, in the knowledge that to a certain extent individuals and societies themselves establish their identities.[30]

IV

The [preceding] two excurses were meant to make plausible the search for homologous structures of consciousness in ego development and social evolution in two areas [viz. world views and collective identities] that are not nearly as well researched as the structures of legal and moral representations. All three complexes lead back to structures of linguistically established intersubjectivity. (a) Law and morality serve to regulate action conflicts consensually and thus to maintain an endangered intersubjectivity of understanding among speaking and acting subjects. (b) The demarcation of different universal object domains —one of which appears in the propositional attitude of the observer as external nature, a second in the performative attitude of the participant in interaction as normative social reality, and a third in the expressive attitude of one who expresses an intention as his own subjective nature—makes possible the differentiation (and if necessary thematization) of those validity claims (truth, rightness, truthfulness) that we implicitly tie to all speech actions. (c) Finally, the construction of personal and corresponding collective identities is a necessary presupposition for taking on the general communicative roles, which are provided for in every speaking and acting situation and which find their expression in the employment of personal pronouns.

To be sure, the communication theory I have in mind is not yet developed to a point at which we could adequately analyze the symbolic structures that underlie law and morality, an intersubjectively constituted world, and the identities of persons and collectives. And we are really far from being able to provide convincing reconstructions of the patterns of development of these structures in the spheres of ontogenesis and species history. The very concept of a developmental logic requires additional sharpening before we can say formally what it means to describe the direction of development in ontogenesis and in the history of the species by means of such concepts as universalization and individualization, decentration, autonomization, and becoming reflective. If I stick to this theme in spite of the (for the time being) unsatisfactory degree of explication, it is because I am convinced that normative structures do not simply follow the path of development of reproductive processes and do not simply respond to the pattern of system problems, but that they have instead an *internal history*. In earlier investigations I have tried to argue that holistic concepts such as productive activity and *Praxis* have to be resolved into the basic concepts of communicative action and purposive rational action in order to avoid confusing the two rationalization processes that determine social evolution;[31] the rationalization of action takes effect not only on productive forces but also, and independently, on normative structures.[32]

Purposive-rational actions can be regarded under two different aspects—the empirical efficiency of technical means and the consistency of choice between suitable means. Actions and action systems can be rationalized in both respects. The rationality of means requires technically utilizable, empirical knowledge; the rationality of decisions requires the explication and inner consistency of value systems and decision maxims, as well as the correct derivation of acts of choice. I shall speak of strategic action in the case of competing opponents insofar as they are determined by the intention of influencing each other's decisions in a purposive-rational way, that is, in a way oriented only to each's own success.

In contexts of social action, the rationalization of means and the choice of means signifies a *heightening of productive forces,* that is, a socially significant implementation of knowledge, with the help of which we can improve the technical outfitting, organizational deployment, and qualifications of available labor power. Marx saw in this process the motor of social development. It is of course necessary to distinguish more precisely among (a) the rationality structures and (if appropriate) developmental logic of the knowledge that can be transposed into technologies, into strategies or organizations, and into qualifications; (b) the mechanisms that can explain the acquisition of this knowledge, the corresponding learning processes; and (c) the boundary conditions under which available knowledge can be implemented in a socially significant way. Only these three complexes of conditions together can explain rationalization processes in the sense of the development of productive forces. However, there is now the further question of whether *other* rationalization processes are just as important or even more important for the explanation of social evolution. In addition to the development of the forces of production, Marx regarded social movements as important. But in conceiving of the organized struggle of oppressed classes as itself a productive force, he established between the two motors of social development—technical-organizational progress on the one hand and class struggle on the other—a confusing, in any event an inadequately analyzed, connection.

In contradistinction to purposive-rational action, *communicative action* is, among other things, oriented to observing intersubjectively valid norms that link reciprocal expectations. In communicative action, the validity basis of speech is presupposed. The universal validity claims (truth, rightness, truthfulness), which participants at least implicitly raise and reciprocally recognize, make possible the consensus that carries action in common. In strategic action, this background consensus is lacking; the truthfulness of expressed intentions is not expected, and the norm-conformity of an utterance (or the rightness of the norm itself) is presupposed in a different sense than in communicative action—namely, contingently. One who repeatedly makes senseless moves in playing chess disqualifies himself as a chess player; and one who follows rules other than those constitutive of chess is not playing chess. Strategic action remains indifferent with respect to its motivational conditions, whereas the consensual presuppositions of communicative action can secure motivations. Thus strategic actions must be institutionalized, that is, embedded in intersubjectively binding norms that guarantee the fulfillment of the motivational conditions. Even then we can distinguish the aspect of purposive-rational action—in Parsons' terminology, the task aspect—from the framework of normatively guided communicative action. In purposive-rational action it is supposed only that each subject is following preferences and decision maxims that he has determined for himself—that is, monologically, regardless of whether or not he agrees therein with other subjects. When, therefore, a strategic action system (e.g., war) makes it necessary for several subjects to agree in certain preferences (and to the extent that this agreement is not guaranteed in fact by the interest situations), purposive-rational action has somehow to be bound or institutionalized (e.g., in the framework of the Hague Convention). Institutionalization again means the organization of consensual action resting on intersubjectively recognized validity claims.

Communicative action can be rationalized neither under the technical aspect of the means selected nor under the strategic aspect of the selection of means but only under the moral-practical aspect of the responsibility of the acting subject and the justifiability of the action norm. Whereas the rationalization of purposive-rational action depends on the accumulation of true (empirically or analytically true) knowledge, the rationalizable aspect of communicative action has nothing to do with propositional truth; but it has everything to do with the truthfulness of intentional expressions and with the rightness of norms. The rationality of action oriented to reaching understanding is measured against:

a. Whether a subject truthfully expresses his intentions in his actions (or whether he deceives himself and others because the norm of action is so little in accord with his needs that conflicts arise that have to be defended against unconsciously, through setting up internal barriers to communication).

b. Whether the validity claims connected with norms of action, and recognized in fact, are legitimate (or whether the existing normative context does not express generalizable or compromisable interests, and thus can be stabilized in its de facto validity only so long as those affected can be prevented by inconspicuous restrictions on communication from discursively examining the normative validity claim).

Rationalization here means extirpating those relations of force that are inconspicuously set in the very structures of communication and that prevent conscious settlement of conflicts, and consensual regulation of conflicts, by means of intrapsychic as well as interpersonal communicative barriers.[33] Rationalization means overcoming such systematically distorted communication in which the action-supporting consensus concerning the reciprocally raised validity claims—especially the consensus concerning the truthfulness of intentional expressions and the rightness of underlying norms—can be sustained in appearance only, that is, counterfactually. The stages of law and morality, of ego demarcations and world-views, of individual and collective identity formations, are stages in this process. Their progress cannot be measured against the choice of correct strategies, but rather against the intersubjectivity of understanding achieved without force, that is, against the expansion of the domain of consensual action together with the re-establishment of undistorted communication.

The categorial distinction between purposive-rational and communicative action thus permits us to separate the aspects under which action can be rationalized. As learning processes take place not only in the dimension of objectivating thought but also in the dimension of moral-practical insight, the rationalization of action is deposited not only in forces of production, but also—mediated through the dynamics of social movements—in forms of social integration.[34] Rationality structures are embodied not only in amplifications of purposive-rational action—that is, in technologies, strategies, organizations, and qualifications—but also in mediations of communicative action—in the mechanisms for regulating conflict, in world views, and in identity formations. I would even defend the thesis that the development of these normative structures is the pacemaker of social evolution, for new principles of social organization mean new forms of social integration; and the latter, in turn, first make it possible to implement available productive forces or to generate new ones, as well as making possible a heightening of social complexity.

The prominent place I have given to normative structures in the framework of social evolution could lead to several misunderstandings—one, that the dynamics of species history is to be explained through an internal history of spirit; and another, that a developmental logic is once again to take the place of historical contingencies. Behind the first misunderstanding lies the suspicion that I am quietly dropping the materialist assumptions regarding the motor of social development; the second suspects another logification of history—and philosophical mystifications instead of empirical-scientific analysis. As indicated, I consider these to be misunderstandings.

Let us assume for a moment that developmental patterns for the normative structures of a certain society can be reconstructed and corroborated. (I am not talking here of any arbitrarily selected classification of stages but of developmental logics in Piaget's sense, which must satisfy rather improbable conditions.) [35] Such rationally reconstructible patterns then represent *rules for possible problemsolving,* that is, only formal restrictions and not mechanisms that could explain individual problemsolving processes, not to mention the acquisition of general problemsolving abilities. The *learning mechanisms* have to be sought first on the psychological level. If that succeeds, with the help of cognitive developmental psychology, there is need for additional empirical assumptions that might explain sociologically how individual learning processes find their way into a society's collectively accessible store of knowledge. *Individually acquired learning abilities* and information must be latently available in world views before they can be used in a socially significant way, that is, before they can be transposed into *societal learning processes.*

Since the cognitive development of the individual takes place under social boundary conditions, there is a circular process between societal and individual learning. To be sure, one could argue for a primacy of social over individual structures of consciousness on the grounds that the rationality structures embodied in the family have first to be absorbed by the child in the development of his interactive competence (as he passes out of the preconventional stage). On the other hand, the initial state of archaic societies—characterized by a conventional kinship organization, a preconventional stage of law, and an egocentric interpretive system—could itself be changed only by constructive learning on the part of socialized individuals. It is only in a derivative sense that societies "learn." I shall assume two series of initial conditions for evolutionary learning processes of society: on the one hand, unresolved system problems that represent challenges;

on the other, new levels of learning that have already been achieved in world views and are latently available but not yet incorporated into action systems and thus remain institutionally inoperative.

System problems express themselves as disturbances of the reproduction process of a society that is normatively fixed in its identity. Whether problems arise which overload the adaptive capacity of a society is a contingent matter; when problems of this type do arise, the reproduction of the society is placed in question—unless it takes up the evolutionary challenge and alters the established form of social integration that limits the employment and development of resources. *Whether* this alteration—which Marx describes as an overthrow of the relations of production—is actually possible, and *how* it is developmental-logically possible, cannot be read off the system problems; it is rather a question of access to a new learning level. The solution to the problems producing the crisis requires (a) attempts to loosen up the existing form of social integration by embodying in new institutions the rationality structures already developed in world views, and (b) a milieu favorable to the stabilization of successful attempts. Every economic advance can be characterized in terms of institutions in which rationality structures of the next higher stage of development are embodied—for example, the royal courts of justice, which, early in the development of civilization, permitted administration of justice at the *conventional* level of moral consciousness; or the capitalist firm, rational administration of the state, and bourgeois norms of civil law, which, at the beginning of the modern period, organized morally neutral domains of strategic action according to *universalistic* principles. Previously sociologists talked only of an "institutionalization of values," through which certain value orientations receive binding force for actors. When I now attempt to grasp evolutionary learning processes with the aid of the concept of "the institutional embodiment of structures of rationality," it is no longer a question of making orienting *contents* binding but of opening up *structural possibilities for the rationalization of action.*

Looking now at this explanatory strategy (which has proven itself in Klaus Eder's investigation of the rise of societies organized around a state), we can see that the objections mentioned above are pointless. The analysis of developmental dynamics is "materialist" insofar as it makes reference to crisis-producing systems problems in the domain of production and reproduction; and it remains "historically" oriented insofar as it has to seek the causes of evolutionary changes in the whole range of those contingent circumstances under which (a) new structures are acquired in the individual consciousness and transposed into structures of world views; (b) system problems arise, which overload the steering capacity of a society; (c) the institutional embodiment of new rationality structures can be tried and stabilized; and (d) the new latitude for the mobilization of resources can be utilized. Only after rationalization processes (which require explanations that are both historical and materialist) have been historically completed can we specify the patterns of development for the normative structures of society. These developmental logics betoken the independence—and to this extent the internal history— of the spirit.

V

Finally I would like to take up two objections that might be directed against my declared intention to take historical materialism as my starting point. In the first place, the investigation of the capitalist accumulation process, on which Marx concentrated above all, hardly plays a role in the reformulation of the basic assumptions regarding social evolution. Instead there are unmistakable borrowings from structuralism and functionalism. Why then insist any longer on the *Marxist* theoretical tradition? Furthermore, why should one pursue historical materialism at all, if the intention of orienting action would be better served by an analysis of the *contemporary* formation of society?

In reference to the first question, the anatomy of bourgeois society is a key to the anatomy of premodern societies; to this extent the analysis of capitalism provides an excellent entry into the theory of social evolution. The general concept of a principle of social organization can be discerned in capitalist societies because there, with the relation of wage labor and capital, the class structure appeared for the first time in a pure, namely, an economic, form. Moreover, the model of the generation of crises that threaten [a society's] existence can be developed in connection with the accumulation process because, with the capitalist economic system, for the first time a system was differentiated that had the specific function of dealing with the tasks of material reproduction. Finally, the mechanism of legitimating domination can be grasped in bourgeois ideologies because there, for the first time, universalistic value systems incompatible with class structures were made unreservedly explicit and were argumentatively grounded. To this extent, the constitutive features of this mode of production are also instructive for social formations in earlier stages, but from this one cannot derive a demand that "the logic of capital" be utilized as the key to the logic of social evolution. For the way in which disturbances of the reproduction process appear in capitalist economic systems cannot be generalized and transposed to other social formations. Moreover, we cannot find in the logic of the rise of system problems the logic that the social system will follow if it responds to such an evolutionary challenge. If a socialist organization of society *were* the adequate response to crisis-ridden developments in capitalist society, it could not be deduced from any "determination of the form" of the reproductive process, but would have to be explained in terms of processes of democratization; that is, in terms of the penetration of universalistic structures into action domains, which—the purposive-rationality of the choice of means notwithstanding—were previously reserved to the private autonomous setting of ends.

As regards taking structuralist points of view into consideration, I readily admit to having learned something from Marxists like Godelier.[36] They have rethought the base-superstructure relationship and conceptualized it in such a way that the proper contribution of normative structures can be saved from a reductionistic short-circuiting. To be sure, the concepts of objective spirit and of culture developed in the Hegelian-Marxist tradition from Lukács to Adorno are not in need of this reformulation. The stimulus that encouraged me to bring normative structures into a developmental-logical problematic came from the *genetic structuralism* of Jean Piaget as well, thus from a conception that has overcome the traditional structuralist front against evolution-ism and that has assimilated motifs of the theory of knowledge from Kant to Peirce. (Lucien Goldmann very early recognized the significance of Piaget's work for Marxist theory.)[37]

Functionalism followed a path that led beyond the cultural anthropology of the thirties and forties and again made possible a connection with the developmental theories of the nineteenth century. Talcott Parsons' neoevolutionism applies the conceptual apparatus of general systems theory to societies and to the struc-tural change of social systems. Functionalist analysis brings social evolution under the viewpoint of the heightening of complexity. In several essays, I have tried to show that this approach comes up short. Functionalism explains evolutionary advances by corre-lating functionally equivalent solutions to system problems. It thus steers away from the evolutionary learning processes that could alone have explanatory power. This explanatory gap is quite evident to a past master of functionalism like N. Eisenstadt —it can be filled with a theory of social movements. If I am not mistaken, A. Tourraine was the first to introduce this element systematically into the theory of social evolution.[38] Naturally, the action orientations that achieve dominance in social movements are, for their part, structured by cultural traditions. If one con-ceives of social movements as learning processes through which latently available structures of rationality are transposed into social practice[39]—so that in the end they find an institutional embodiment—there is the further task of identifying the rational-ization potential of traditions.

Nevertheless, systems theory offers useful instruments for analyzing the initial conditions of evolutionary innovations, namely, the appearance of system problems that overload a struc-turally limited steering capacity and trigger crises that endanger the system's continued existence. Claus Offe has shown how systems-theoretic concepts and hypotheses can be used precisely for the analysis of crises[40]—at least if one connects systems theory and action theory. But then we need an equivalent for the rules of translation that Marx provided (in the form of a theory of value) for the connection between circulation processes and class structure, between value relations and power relations.

[In reference to the second question] it is another question whether in Marx historical materialism did not have the rather incidental role of merely complementing the analysis of capitalism with a backward glance at precapitalist societies and whether the analysis of the contemporary formation of society ought not to stand on its own feet. Marx was concerned to identify and to explain the developments that showed the structural limitation of adaptive capacity and made it possible to ground the *practical* necessity for a change in the organizational principle of society. If it is true that historical materialism cannot contribute much to these questions, then the interest in questions of historical materialism has to arouse the suspicion of escapism. But I am of the opinion that Marx already understood historical materialism as a comprehensive theory of social evolution and regarded the theory of capitalism as one of its subparts. Leaving Marx's view to one side, the theory of social evolution has a precisely specifiable, systematic significance for an analysis of the present that inquires about the exhaustion of the innovative and adaptive potential of existing social structures.

Assumptions about the organizational principle of society and about learning capacities and ranges of possible structural variation cannot be clearly checked empirically before historical developments have put the critical survival limits to the test. Evolutionarily oriented analyses of the present are alway handicapped because they cannot view their object retrospectively. For that reason, theories of this type, whether Marxist or non-Marxist, are forced to monitor their assumptions—assumptions that already underlie the delimitation and description of the object—on an instructive theory of social development. Characterizations of society as industrial, postindustrial, technological, scientific, capitalist, late-capitalist, state-monopolistic, state-capitalist, totally administered, tertiary, modern, postmodern, and so on, stem from just as many developmental models connecting the contemporary formation of society with earlier ones. In this regard, historical materialism can take on the task of determining the organizational principle of contemporary society from the perspective of the origin of this social formation—for example, with statements about the systems problems in the face of which traditional societies failed and about the innovations with which modern bourgeois society met the evolutionary challenges. I would like to illustrate with two examples the *kind* of question that, in my view, requires that we again take up historical materialism.

In an in-house working paper, R. Funke contrasted two theoretical approaches to the analysis of developed capitalist societies: theories of "still-capitalism," which start with the idea that the capitalist organizational principle is already limited in its effectiveness by a new political principle of organization that has to be further specified; and, on the other hand, theories of "yet-to-be-accomplished-capitalism," which start with the idea that capitalism is still being established, that it is still in the process of clearing away the remains of tradition from quasi-natural social relations and infrastructures and of integrating them into the accumulation process and the commodity form. From evolutionary perspectives the significance of the same facts is rather different according to whether they are supposed to support a view of the state springing into functional gaps in the market as a substitute or one of the administrative establishment of the commodity form for previously quasi-natural social relations. The same crisis phenomena signify in one perspective the exhaustion of capitalistically limited ranges of variation and, in the other, the dilemma of a capitalism that has to transform inherited social relations and infrastructures without being able to regenerate their stabilizing powers. If, as I shall assume for the sake of my argument, the rival interpretations could explain the available data more or less equally well, how can we decide between them?

If we had a theory of social evolution that explained the transition to the modern world as the rise of a new and, moreover, well-defined, organizational principle of society, there would be a possibility of examining which of the two competing approaches was more compatible with this explanation, for these two different interpretations advance different organizational principles for capitalist development. According to the first version, the principle of organization consists in a complementary relationship between a nonproductive state and a depoliticized economic system. The latter is organized through markets—that is, in accordance with general and abstract rules—as a domain of decentralized decisions of strategically acting private subjects; whereas the state guarantees the presuppositions for the continued existence of an economy differentiated from its domain of sovereignty, and thereby excludes itself from the process of production, while at the same time—as a state based on taxation—making itself dependent on it.[41] According to the other version, the principle of organization consists in the relationship between capital and wage labor; the state however (somewhat ex machina) has to function as the agent for establishing this principle in an initially alien environment. In the one case, the depoliticization of a process of production that is *in fact* controlled through markets is constitutive for the mode of production; in the other case, it is the state-enforced expansion of an interaction network that is *formally* regulated through exchange relations that is constitutive.

Another example that can elucidate the systematic importance of historical materialism is the question of classifying bureaucratic-socialist societies. Here I cannot even run through the most important interpretations that have been offered for this ambiguous complex. Instead I shall indicate a perspective from which the different interpretations can be roughly classified. In one version, societies of the bureaucratic-socialist type have, in comparison to developed capitalist societies, reached a higher stage of evolution. In the other version, it is a question of two variants of the same stage of development—that is, different historical expressions of the same principle of organization. The second version is represented not only in the trivial form of (largely invalidated) convergence theses but also by theoreticians who— Adorno, for instance—by no means play down the system-specific differences in the mode of production but yet (with Max Weber) attribute a weight of its own to the autonomization of instrumental rationality.[42] If this version could be corroborated, the complementary relationship of state and economy that is characteristic of modern societies would have to be grasped quite abstractly; for then the relation of the state based on taxation to the capitalist economy, which is constitutive for bourgeois society, would represent only one of its possible realizations. On this presupposition critical developments do not automatically have to count as indicators for the exhaustion of structurally limited steering capacities; under certain circumstances they are also signs that the rationality structures that *became accessible* in the modern age have *not yet* been exhausted and that they allow for a comprehensive institutional embodiment in the form of extensive processes of democratization.[43]

These examples have a strong admixture of speculation; for just that reason they serve to illustrate a class of questions that are important enough to *deserve* more rigorous argumentation and that *could* be dealt with in a more rigorously argumentative manner if we did not renounce from the outset the level of analysis of historical materialism or of a theory of social evolution satisfying its claims. An evolutionarily oriented analysis of the present that does renounce it has to proceed dogmatically with the kind of question for which I have supplied examples.[44]

4 Toward a Reconstruction of Historical Materialism

Only twice did Marx express himself connectedly and fundamentally on the materialist conception of history;[1] otherwise he used this theoretical framework, in the role of historian, to interpret particular historical situations or developments—unsurpassedly in *The Eighteenth Brumaire of Louis Bonaparte*. Engels characterized historical materialism as a guide and a method.[2] This could create the impression that Marx and Engels saw this doctrine as no more than a heuristic that helped to structure a (now-as-before) narrative presentation of history with systematic intent. But historical materialism was not understood in this way—either by Marx and Engels or by Marxist theoreticians or in the history of the labor movement. I shall not, therefore, treat it as a heuristic but as a theory, indeed as a theory of social evolution that, owing to its reflective status, is also informative for purposes of political action and can under certain circumstances be connected with the theory and strategy of revolution. The theory of capitalist development that Marx worked out in the *Grundrisse* and in *Capital* fits into historical materialism as a *subtheory*.

In 1938 Stalin codified historical materialism in a way that has proven of great consequence;[3] the historical-materialist research since undertaken has remained largely bound to this theoretical

framework.[4] The version set down by Stalin needs to be reconstructed. My attempt to do so is also intended to further the
critical appropriation of competing approaches—above all of
neoevolutionism and of structuralism. Of course, I shall be able
to make plausible only a few viewpoints from which such a reconstruction might be attempted with some hope of success.

I would like first to introduce and consider critically some basic
concepts and assumptions of historical materialism; I shall then
point out certain difficulties that arise in applying its hypotheses
and advance and illustrate an (abstract) proposal for resolving
them; finally, I shall see what can be learned from competing
approaches.

I

To begin I shall examine the concepts of *social labor* and *history
of the species*, as well as three fundamental assumptions of historical materialism.

1. *Socially organized labor* is the specific way in which humans,
in contradistinction to animals, reproduce their lives.

Man can be distinguished from the animal by consciousness, religion,
or anything else you please. He begins to distinguish himself from the
animal the moment he begins to *produce* his means of subsistence, a
step required by his physical organization. By producing food, man
indirectly produces his material life itself.[5]

At a level of description that is unspecific in regard to the human
mode of life, the exchange between the organism and its environment can be investigated in the physiological terms of ma-
terial-exchange processes. But to grasp what is specific to the
human mode of life, one must describe the relation between
organism and environment at the level of labor processes. From
the physical aspect the latter signify the expenditure of human
energy and the transfer of energies in the economy of external
nature; but what is decisive is the sociological aspect of the goal-
directed transformation of material according to *rules of instrumental action*.[6]

Of course, under "production" Marx understands not only the

instrumental actions of a single individual, but also the *social cooperation* of different individuals:

The production of life, of one's own life in labor, and of another in procreation, now appears as a double relationship: on the one hand as a natural relationship, on the other as a social one. The latter is social in the sense that individuals co-operate, no matter under what conditions, in what manner, and for what purpose. Consequently a certain mode of production or industrial stage is always combined with a certain mode of co-operation or social stage, and this mode of co-operation is itself a "productive force." We observe in addition that the multitude of productive forces accessible to men determines the nature of society and that the "history of mankind" must always be studied and treated in relation to the history of industry and exchange.[7]

The instrumental actions of different individuals are coordinated in a purposive-rational way, that is, with a view to the goal of production. The *rules of strategic action,* in accord with which cooperation comes about, are a necessary component of the labor process.

Means of subsistence are produced only to be consumed. The distribution of the product of labor is, like the labor itself, socially organized. In the case of rules of distribution, the concern is not with processing material or with the suitably coordinated application of means, but with the systematic connection of reciprocal expectations or interests. Thus the distribution of products requires rules of interaction that can be set intersubjectively at the level of linguistic understanding, detached from the individual case, and made permanent as recognized norms or *rules of communicative action.*

We call a system that socially regulates labor and distribution an *economy.* According to Marx, then, the economic form of reproducing life is characteristic of the human stage of development.

The concept of social labor as the *form of reproduction of human life* has a number of connotations. It is critical of the most basic assumptions of the modern philosophy of the subject or reflection. The statement—"As individuals express their life, so they are. What they are, therefore, coincides with *what* they produce and *how* they produce" [8]—can be understood, according to the first of the *Theses on Feuerbach,* in the sense of an epistemologically oriented *pragmatism,* that is, as a critique of a phenomenalism of any sort, empiricist or rationalist, which understands the knowing subject as a passive, self-contained consciousness. The same statement has *materialist* connotations as well; it is directed equally against theoretical and practical idealism, which assert the primacy of the spirit over nature and that of the idea over the interest. Or consider the statement: "But the essence of man is no abstraction inhering in each single individual. In its actuality it is the ensemble of social relationships." [9] Here Marx, schooled in the Hegelian concept of objective spirit, declares war on the methodological individualism of the bourgeois social sciences and on the practical individualism of English and French moral philosophy; both set forth the acting subject as an isolated monad.

In the present context we are naturally interested in the question, whether the concept of social labor adequately characterizes the form of reproduction of human life. Thus we must specify more exactly what we wish to understand by "human mode of life." In the last generation anthropology has gained new knowledge about the long (more than four million years) phase during which the development from primates to humans, that is, the process of hominization, took place; beginning with a postulated common ancestor of chimpanzees and humans, the evolution proceeded through homo erectus to homo sapiens. This hominization was determined by the *cooperation of organic and cultural mechanisms of development.* On the one hand, during this period of anthropogenesis, there were changes—based on a long series of mutations—in the size of the brain and in important morphological features. On the other hand, the environments from which the pressure for selection proceeded were no longer determined solely by natural ecology, but through the active, adaptive accomplishments of hunting bands of hominids. Only at the threshold to homo sapiens did this mixed organic-cultural form of evolution give way to an exclusively *social evolution.* The natural mechanism of evolution came to a standstill. No new species arose. Instead, the exogamy that was the basis for the societization of homo sapiens resulted in a broad, intraspecific dispersion and mixture of the genetic inheritance. This internal differentiation was the natural basis for a cultural diversification evidenced in a multiplicity of social learning processes. It is therefore advisable to demarcate the sociocultural stage of development—at which alone social evolution takes place (i.e., society is caught up in evolution)—from not only the primate stage—at which there is a still exclusively natural evolution (i.e., the species are caught up in evolution)—but also from the hominid stage—at which the two evolutionary mechanisms are working together, the evolution of the brain being the most important single variable. [10]

2. If we examine the concept of social labor in the light of more recent anthropological findings, it becomes evident that it cuts too deeply into the evolutionary scale; not only humans but hominids too were distinguished from the anthropoid apes in that they converted to reproduction through social labor and developed an economy. The adult males formed hunting bands, which (a) made use of weapons and tools (technology), (b) cooperated through a division of labor (cooperative organization), and (c) distributed the prey within the collective (rules of distribution). The making of the means of production and the social organization of labor, as well as of the distribution of its products, fulfilled the conditions for an economic form of reproducing life.

The society of hominids is more difficult to reconstruct than their mode of production. It is not clear how far beyond interactions mediated by gestures—already found among primates—their system of communication progressed. The conjecture is that they possessed a *language* of gestures and a system of *signal calls*.[11] In any event, cooperative big-game hunting requires reaching understanding about experiences, so that we have to assume a protolanguage, which at least paved the way for the systematic connection of cognitive accomplishments, affective expressions, and interpersonal relations that was so important for hominization. The division of labor in the hominid groups presumably led to a development of two subsystems: on the one hand, the adult males, who were together in egalitarian hunting bands and occupied, on the whole, a dominant position; on the other hand, the females, who gathered fruit and lived together with their young, for whom they cared. In comparison to primate societies, the strategic forms of cooperation and the rules of distribution were new; both innovations were directly connected with the establishment of the *first mode of production,* the cooperative hunt.

Thus the Marxian concept of social labor is suitable for delimiting the mode of life of the hominids from that of the primates; but it does not capture the specifically human reproduction of life. Not hominids, but humans were the first to break up the social structure that arose with the vertebrates—the one-dimensional rank ordering in which every animal was transitively assigned one and only one status. Among chimpanzees and baboons this status system controlled the rather aggressive relations between adult males, sexual relations between male and female, and social relations between the old and the young. A familylike relationship existed only between the mother and her young, and between siblings. Incest between mothers and growing sons was not permitted;[12] there was no corresponding incest barrier between fathers and daughters, because the father role did not exist. Even hominid societies converted to the basis of social labor did not yet know a family structure. We can, of course, imagine how the family might have emerged. The mode of production of the socially organized hunt created a system problem that was resolved by the familialization of the male (Count),[13] that is, by the introduction of a kinship system based on exogamy. The male society of the hunting band became independent of the plant-gathering females and the young, both of whom remained behind during hunting expeditions. With this differentiation, linked to the division of labor, there arose a new need for integration, namely, the need for a controlled exchange between the two subsystems. But the hominids apparently had at their disposal only the pattern of status-dependent sexual relations. This pattern was not equal to the new need for integration, the less so, the more the status order of the primates was further undermined by forces pushing in the direction of egalitarian relations within the hunting band. Only a family system based on marriage and regulated descent permitted the adult male member to link—via the father role—a status in the male system of the hunting band with a status in the female and child system, and thus (1) integrate functions of social labor with functions of nurture of the young, and, moreover, (2) coordinate functions of male hunting with those of female gathering.

3. We can speak of the reproduction of *human* life, with homo sapiens, only when the economy of the hunt is supplemented by a familial social structure. This process lasted several million years; it represented an important replacement of the animal status system, which among the anthropoid apes was already based on symbolically mediated interaction (in Mead's sense) by a system of social norms that presupposed *language*. The rank order of the primates was one-dimensional; every individual could occupy one and only one—that is, in all functional domains the same—status. Only when the same individual could unify various status positions and different individuals could occupy the same status was a socially regulated exchange between functionally specified subsystems possible. The animal status system was based on the status occupant's capacity to threaten, that is, on power as an attribute of personality. By contrast, social role systems are based on the intersubjective recognition of normed expectations of behavior and not on respect for the possibilities of sanction situationally available to a role occupant because of peculiarities of his personality structure. This change means a *moralization of motives for action*. Social roles can conditionally link two different behavioral expectations in such a way that a system of reciprocal motivation is formed. Alter can count on ego fulfilling his (alter's) expectations because ego is counting on alter fulfilling his (ego's) expectations. Through social roles social influence on the motives of the other can be made independent of accidental, situational contexts, and motive formation can be brought into the symbolic world of interaction. For this to occur, however, three conditions must be met:

a. Social roles presuppose not only that participants in interaction can assume the perspective of other participants (which is already the case in symbolically mediated interaction), but that they can also exchange the perspective of the participant for that of the observer. Participants must be able to adopt, in regard to themselves and others, the perspective of an observer, from which they view the system of their expectations and actions from the outside, as it were. Otherwise they could not conditionally link their reciprocal expectations and make them, as a system, the basis of their own action.[14]

b. Social roles can be constituted only if the participants in interaction possess a *temporal horizon* that extends beyond the immediately actual consequences of action. Otherwise spatially, temporally, and materially differentiated expectations of behavior could not be linked with one another in a single social role. Burial rites are a sign that living together as a family induced a categorically expanded consciousness of time.[15]

c. Social roles have to be connected with mechanisms of sanction if they are to control the action motives of participants. Since [in the first human societies] the possibility of sanction was no longer (as in primate societies) covered by the accidental qualities of concrete reference persons and was not yet (as in civilizations) covered by the means of power of political domination, it could consist only in ambivalently cathected interpretations of established norms. As can be seen in the way that taboos function, interpretive patterns tied to social roles reworked the feeling ambivalence, which must have resulted from dedifferentiating the drive system, into the consciousness of normative validity, that is, into readiness to respect established norms.[16]

For a number of reasons these three conditions could not be met before language was fully developed. We can assume that the developments that led to the specifically human form of reproducing life—and thus to the initial state of social evolution —first took place in the structures of labor and language. *Labor and language are older than man and society.* For the basic anthropological concepts of historical materialism this might imply the following:

a. The concept of social labor is fundamental, because the evolutionary achievement of socially organized labor and distribution obviously precedes the emergence of developed linguistic communication, and this in turn precedes the development of social role systems.

b. The specifically human mode of life, however, can be adequately described only if we combine the concept of social labor with that of the familial principle of organization.

c. The structures of role behavior mark a new stage of development in relation to the structures of social labor; rules of communicative action, that is, intersubjectively valid and ritually secured norms of action, cannot be reduced to rules of instrumental or strategic action.

d. Production and socialization, social labor and care for the young, are equally important for the reproduction of the species; thus the familial social structure, which controls both—the integration of external as well as of internal nature—is fundamental.[17]

II

Marx links the concept of social labor with that of the *history of the species*. This phrase is intended in the first place to signal the materialist message that in the case of a single species natural evolution was continued by other means, namely, through the productive activity of the socialized individuals themselves. In sustaining their lives through social labor, men produce at the same time the material relations of life; they produce their society and the historical process in which individuals change along with their societies. The key to the reconstruction of the history of the species is provided by the concept of a *mode of production*. Marx conceives of history as a discrete series of modes of production, which, in its developmental-logical order, reveals the direction of social evolution. Let us recall the most important definitions.

A *mode of production* is characterized by a specific state of development of productive forces and by specific forms of social intercourse, that is, relations of production. The *forces of production* consist of (1) the labor power of those engaged in production, the producers; (2) technically useful knowledge insofar as it can be converted into instruments of labor that heighten productivity, that is, into technologies of production; (3) organizational knowledge insofar as it is applied to set labor power efficiently into motion, to qualify labor power, and to effectively coordinate the cooperation of laborers in accord with the division of labor (mobilization, qualification, and organization of labor power). Productive forces determine the degree of possible control over natural processes. On the other hand, the *relations of production* are those institutions and social mechanisms that determine the way in which (at a given stage of productive forces) labor power is combined with the available means of production. Regulation of access to the means of production, the way in which socially employed labor power is controlled, also determines indirectly the distribution of socially produced wealth. The relations of production express the distribution of social power; with the distributional pattern of socially recognized opportunities for need satisfaction, they prejudge the *interest structure* of a society. Historical materialism proceeds from the assumption that productive forces and productive relations do not vary independently, but form structures that (a) correspond with one another and (b) yield a finite number of structurally analogous stages of development, so that (c) there results a series of modes of production that are to be ordered in a developmental logic. (The handmill produces a society of feudal lords, the steam mill a society of industrial capitalists.)[18]

In the orthodox version, five modes of production are distinguished: (1) the primitive communal mode of bands and tribes prior to civilization; (2) the ancient mode based on slaveholding; (3) the feudal; (4) the capitalist; and finally (5) the socialist modes of production. The discussion of how the ancient Orient and the ancient Americas were to be ordered in this historical development led to the insertion of (6) an Asiatic mode of production.[19] These six modes of production are supposed to mark universal stages of social evolution. From an evolutionary standpoint, every particular *economic structure* can be analyzed in terms of the various modes of production that have entered into a hierarchical combination in a historically concrete society. (A good example of this is Godelier's analysis of the Inca culture at the time of Spanish colonization.)[20]

The *dogmatic version* of the concept of a history of the species shares a number of weaknesses with eighteenth-century designs for a philosophy of history. The course of previous world history, which evidences a sequence of five or six modes of production, sets down the *unilinear, necessary, uninterrupted, and progressive development of a macrosubject*. I should like to oppose to this model of species history a weaker version, which is not open to the familiar criticisms of the objectivism of philosophy of history.[21]

a. Historical materialism does not need to assume a *species-subject* that undergoes evolution. The bearers of evolution are rather societies and the acting subjects integrated into them; social evolution can be discerned in those structures that are replaced by more comprehensive structures in accord with a pattern that is to be rationally reconstructed. In the course of this structure-forming process, societies and individuals, together with their ego and group identities, undergo change.[22] Even if social evolution should point in the direction of unified individuals consciously influencing the course of their own evolution, there would not arise any large-scale subjects, but at most self-established, higher-level, intersubjective commonalities. (The specification of the concept of development is another question: in what sense can one conceive the rise of new structures as a movement?—only the empirical substrates are in motion.)[23]

b. If we separate the logic from the dynamics of development —that is, the rationally reconstructible *pattern* of a hierarchy of more and more comprehensive structures from the *processes* through which the empirical substrates develop—then we need require of history neither unilinearity nor necessity, neither continuity nor irreversibility. We certainly do reckon with anthropologically deep-seated general structures, which were formed in the phase of hominization and which lay down the initial state of social evolution; these structures presumably arose to the extent that the cognitive and motivational potential of the anthropoid apes was transformed and reorganized under conditions of linguistic communication. These basic structures correspond, possibly, to the structures of consciousness that children today normally master between their fourth and seventh years, as soon as their cognitive, linguistic, and interactive abilities are integrated with one another.

Such structures describe the logical space in which more comprehensive structural formations can take shape; whether new structural formations arise at all, and if so, when, depends on *contingent* boundary conditions and on learning processes that can be investigated empirically. The genetic explanation of why a certain society has attained a certain level of development is independent of the structural explanation of how a system behaves—a system that conforms at every given stage to the logic of its acquired structures. Many paths can lead to the same level of development; *unilinear* developments are all the less probable, the more numerous the evolutionary units. Moreover, there is no guarantee of uninterrupted development; rather, it depends on accidental constellations whether a society remains unproductively stuck at the threshold of development or whether it solves its system problems by developing new structures. Finally, *retrogressions* in evolution are possible and in many cases empirically corroborated; of course, a society will not fall back behind a level of development, once it is established, without accompanying phenomena of forced regression; this can be seen, for example, in the case of Fascist Germany. It is not evolutionary processes that are *irreversible* but the structural sequences that a society must run through *if* and *to the extent that* it is involved in evolution.

3. Naturally the most controversial point is the *teleology* that, according to historical materialism, is supposed to be inherent in history. When we speak of evolution, we do in fact mean cumulative processes that exhibit a direction. Neoevolutionism regards *increasing complexity* as an acceptable directional criterion. The more states a system can assume, the more complex the environment with which it can cope and against which it can maintain itself. Marx too attributed great significance to the category of the "social division of labor"; by this he meant processes of system differentiation and of integration of functionally specified subsystems at a higher level, that is, processes that increase the internal complexity—and thereby the adaptive capacity—of a society. However, as a social-evolutionary directional criterion, complexity has a number of disadvantages:

a. Complexity is a multidimensional concept. A society can be complex with respect to size, interdependence, and variability, with respect to achievements of generalization, integration, and respecification. As a result, complexity comparisons can become blurred, and questions of global classification from the viewpoint of complexity undecidable.[24]

b. Moreover, there is no clear relation between complexity and self-maintenance. There are increases in complexity that turn out to be evolutionary dead ends. But without this connection, increases in complexity are unsuitable as directional signs; system complexity is equally ill-suited to be the basis for evolutionary stages of development.

c. The connection between complexity and self-maintenance becomes problematic because societies, unlike organisms, do not have clear-cut boundaries and objectively decidable problems of self-maintenance. The reproduction of societies is not measured in terms of rates of reproduction, that is, possibilities of the physical survival of their members, but in terms of securing a normatively prescribed societal identity, a culturally interpreted "good" or "tolerable" life.[25]

Marx judged social development not by increases in complexity but by the stage of development of productive forces and by the maturity of the forms of social intercourse.[26] The development of productive forces depends on the application of technically useful knowledge; and the basic institutions of a society embody moral-practical knowledge. Progress in these two dimensions is measured against the two universal validity claims we also use to measure the progress of empirical knowledge and of moral-practical insight, namely, the truth of propositions and the rightness of norms. I would like, therefore, to defend the thesis that the criteria of social progress singled out by historical materialism as the development of productive forces and the maturity of forms of social intercourse can be systematically justified. I shall come back to this.

III

Having elucidated the concepts of *social labor* and *history of the species*, I want to look briefly at two basic assumptions of historical materialism: the superstructure theorem and the dialectic of the forces and relations of production.

1. The best-known formulation of the superstructure theorem runs as follows:

In the social production of their existence, men inevitably enter into definite relations of production appropriate to a given stage in the development of their material forces of production. The totality of these relations of production constitutes the economic structure of society, the real foundation, on which arises a legal and political superstructure and to which correspond definite forms of social consciousness. The mode of production of material life conditions the general process of social, political and intellectual life. It is not the consciousness of men that determines their existence, but their social existence that determines their consciousness.[27]

In every society the forces and relations of production form—in accordance with the dominant mode of production—an economic structure by which all other subsystems of the society are determined. For a long time an *economistic version* of this theorem was dominant. On this interpretation every society is divided—in accord with its degree of complexity—into subsystems that can be hierarchically placed in the order: economic sphere, administrative-political sphere, social sphere, cultural sphere. The theorem then states that processes in any higher subsystems are determined, in the sense of causal dependency, by processes in the subsystems below it. A weaker version of this thesis states that lower subsystems place structural limits on developments in systems higher than themselves. Thus the economic system determines "in the final analysis," as Engels puts it, the scope of the developments possible in other subsystems. In Plekhanov we find formulations that support the first interpretation; in Labriola and Max Adler, passages that support the second. Among Hegelian Marxists like Lukács, Korsch, and Adorno, the concept of the social totality excludes a model of levels. The superstructure theorem here posits a kind of concentric dependency of all social appearances on the economic structure, the latter being conceived dialetically as the essence that comes to existence in the observable appearances.

The context in which Marx put forth his theorem makes it clear, however, that the dependency of the superstructure on the base was intended in the first instance only for the critical phase in which a society passes into a new developmental level. It is not some ontological interpretation of society that is intended but the leading role that the economic structure assumes in social evolution. Interestingly Karl Kautsky saw this:

Only in the final analysis is the whole legal, political, ideological apparatus to be regarded as a superstructure over an economic infrastructure. This in no way holds for its individual appearance in history.

The latter—whether of an economic, ideological, or some other type—will act in many respects as infrastructure, in others as superstructure. The Marxian statement about infrastructure and superstructure is unconditionally valid *only for new appearances in history.*[28]

Marx introduced the concept of *base* in order to delimit a domain of problems to which an explanation of evolutionary innovations must make reference. The theorem states that evolutionary innovations only solve those problems that arise in the basic domain of society.

The equation of *base* and *economic structure* could lead to the view that the basic domain always coincides with the economic system. But this is true only of capitalist societies. We have specified the relations of production by means of their function of regulating access to the means of production and thereby indirectly regulating the distribution of social wealth. In primitive societies this function was performed by kinship systems, and in civilizations, by systems of domination. Only in capitalism, when the market, along with its steering function, also assumed the function of stabilizing class relationships, did the relations of production come forth as such and take on an economic form. The theories of postindustrial society even envision a state in which evolutionary primacy would pass from the economic system to the educational and scientific system.[29] Be that as it may, the relations of production can make use of different institutions.[30]

The institutional core around which the relations of production crystallize lays down a specific form of social integration. By *social integration,* I understand, with Durkheim, securing the unity of a social life-world through values and norms. If system problems cannot be solved in accord with the dominant form of social integration, if the latter must itself be revolutionized in order to create latitude for new problem solutions, the identity of society is in danger.

2. Marx sees the mechanism of crisis as follows:

At a certain stage of development, the material productive forces of society come into conflict with the existing relations of production or—this merely expresses the same thing in legal terms—with the property relations within the framework of which they have operated hitherto.

From forms of development of the productive forces, these relations turn into their fetters. Then begins an era of social revolution. The changes in the economic foundation lead sooner or later to the transformation of the whole immense superstructure.[31]

The dialectic of forces and relations of production has often been understood in a *technologistic sense*. The theorem then states that techniques of production necessitate not only certain forms of organizing and mobilizing labor power, but also, through the social organization of labor, the relations of production appropriate to it. The production process is conceived as so unified that relations of production are set up in the very process of deploying the forces of production. In the young Marx, precisely the idealist conceptual apparatus ("the objectification of essential powers in labor") lends support to this idea; in Engels, Plekhanov, Stalin, and others the concept of productive relations "issuing" from productive forces is borne instead by instrumentalist models of action.[32]

We must however separate the level of communicative action from that of the instrumental and strategic action combined in social cooperation. If we take this into account, the theorem can be understood to state that (a) there exists an endogenous learning mechanism that provides for spontaneous growth of technically and organizationally useful knowledge and for its conversion into forces of production; (b) a mode of production is in a state of equilibrium only if there is a structural correspondence between the stages of development of the forces and relations of production; (c) the endogeneously caused development of productive forces makes it possible for structural incompatibilities between the two orders to arise, which (d) bring forth disequilibriums in the given mode of production and must lead to an overthrow of existing relations of production. (Godelier, for example, appropriated the theorem in this *structuralist sense*.[33])

In this formulation too it remains unclear what mechanism could help to explain evolutionary innovations. The postulated learning mechanism explains the growth of a cognitive potential and perhaps also its conversion into technologies and strategies that heighten productivity. It can explain the emergence of system problems that, when the structural dissimilarities between forces and relations of production become too great, threaten the continued existence of the mode of production. But this learning mechanism does not explain how the problems that arise can be solved. The introduction of new forms of social integration—for example, the replacement of the kinship system with the state—requires knowledge of a moral-practical sort and not technically useful knowledge that can be implemented in rules of instrumental and strategic action. It requires not an expansion of our control over external nature but knowledge that can be embodied in structures of interaction—in a word, an extension of the autonomy of society in relation to our own, internal nature.

This can be shown in the example of industrially developed societies. The progress of productive forces has led to a highly differentiated division of labor processes and to a differentiation of the organization of labor within industries. But the cognitive potential that has gone into this "socialization of production" has no structural similarity to the moral-practical consciousness that can support social movements pressing for a revolutionizing of bourgeois society. Thus the advance of industry does not, as the *Communist Manifesto* claims, "replace the isolation of the laborers by their revolutionary combination";[34] rather it replaces an old organization of labor with a new one.

The development of productive forces can then be understood as a problem-generating mechanism that *triggers but does not bring about* the overthrow of relations of production and an evolutionary renewal of the mode of production. But even in this formulation the theorem can hardly be defended. To be sure, we know of a few instances in which system problems arose as a result of an increase in productive forces, overloading the adaptive capacity of societies organized on kinship lines and shattering the primitive communal order—this was apparently the case in Polynesia and South Africa.[35] But the great endogenous, evolutionary advances that led to the first civilizations or to the rise of European capitalism were not conditioned but followed by significant development of productive forces. In these cases the development of productive forces could not have led to an evolutionary challenge.

It is advisable to distinguish between the potential of available knowledge and the implementation of this knowledge. It seems to be the case that the mechanism of not-being-able-not-to-learn (for which Moscovici has supplied intuitive support) again and again provides surpluses that harbor a potential of technical-organizational knowledge utilized only marginally or not at all. When this cognitive potential is drawn upon, it becomes the foundation of structure-forming social divisions of labor (between hunters and gatherers, tillers and breeders, agriculture and city craftsmen, crafts and industry, and so on).[36] The endogenous growth of knowledge is thus a necessary condition of social evolution. But only when a new institutional framework has emerged can the as-yet unresolved system problems be treated with the help of the accumulated cognitive potential; from this there *results* an increase in productive forces. Only in this sense can one defend the statement that a social formation is never destroyed and that new, superior relations of production never replace older ones "before the material conditions for their existence have matured within the framework of the old society."[37]

Our discussion has led to the following, provisional results:

a. The system problems that cannot be solved without evolutionary innovations arise in the basic domain of a society.

b. Each new mode of production means a new form of social integration, which crystallizes around a new institutional core.

c. An endogenous learning mechanism provides for the accumulation of a cognitive potential that can be used for solving crisis-inducing system problems.

d. This knowledge, however, can be implemented to develop the forces of production only when the evolutionary step to a new institutional framework and a new form of social integration has been taken.

It remains an open question, *how* this step is taken. The *descriptive* answer of historical materialism is: through social conflict, struggle, social movements, and political confrontations (which, when they take place under the conditions of a class structure, can be analyzed as class struggles). But only an analytic answer can explain *why* a society takes an evolutionary step and how we are to understand that social struggles under certain conditions lead to a new level of social development. I would like to propose the following answer: the species learns not only in the dimension of technically useful knowledge decisive for the development of productive forces but also in the dimension of moral-practical consciousness decisive for structures of interaction. The rules of communicative action do develop in reaction to changes in the domain of instrumental and strategic action; but in doing so they follow *their own logic*.

IV

The historical-materialist concept of the history of the species calls for reconstructing social development in terms of a *developmental sequence of modes of production*. I would like to indicate a few advantages and difficulties that arise in applying this concept and then put up for discussion a proposed resolution [of the problem].

1. The advantages can be seen through comparison with competing attempts to find viewpoints from which the historical material can be ordered in a developmental logic. Thus there are proposals for periodization based on the principal materials being worked (from stone, bronze, and iron, up to the synthetic products of today) or on the most important energy sources being exploited (from fire, water, and wind, up to atomic and solar energy). But the attempt to discover a developmental pattern in these sequences soon leads to the techniques for making natural resources accessible and for working them. There does, in fact, seem to be a pattern of development to the history of technology.[38] At any rate, technological development accommodates itself to being interpreted *as if* mankind had successively projected the elementary components of the behavioral system of purposive-rational action (which is attached in the first instance to the human organism) onto the level of technical means, and relieved itself of the corresponding functions—at first of the functions of the motor apparatus (legs and hands), then of the functions of the sensory apparatus (eyes, ears, skin) and of the brain.

We can, of course, go beyond the level of the history of technology. In the ontogenetic dimension, Piaget has pointed out a *universal developmental sequence* for cognitive development—from preoperational through concrete-operational to formal-operational thought. The history of technology is probably connected with the great evolutionary advances of society through the *evolution of world views;* and this development might, in turn, be explicable through formal structures of thought for which cognitive psychology has provided a well-examined ontogenetic model, a model that enables us to place these structures in a developmental-logical order.[39]

In any case, since the "neolithic revolution" the great technical discoveries have not brought about new epochs but have merely accompanied them. A history of technology, no matter how rationally reconstructible, is not suited for delimiting social formations. The concept of a mode of production takes into account the fact that the development of productive forces, while certainly an important dimension of social development, *is not decisive* for periodization. Other proposals for periodization are guided by a classification of *forms of cooperation;* and certainly the development from household industries, through their coordination in cottage industry, through factories, national enterprises involving division of labor, up to multinational concerns, does play an important role. But this line of development can be traced only within a single social formation, namely the capitalist; this shows that social evolution cannot be reconstructed in terms of the organization of labor power. The same holds for the development of the *market* (from the household economy, through town and national economies, up to the world economy), or for the *social division of labor* (between hunting and gathering, cultivating and breeding, city crafts and agriculture, agriculture and industry, and so on). These developments increase the complexity of social organization; but it is not written on the face of any of these phenomena, when a new form of organization, a new medium of communication, or a new functional specification means development of productive forces (increased power to dispose of external nature) and when it serves the repression of internal nature and has to be understood as a component of productive relations. For this reason it is more informative to determine the different modes of production directly through relations of production and to analyze changes in the complexity of a society in dependence on its mode of production.[40]

2. There are, of course, also *difficulties* in employing this concept. The decisive point of view here is how access to the means of production is regulated. The state of discussion within historical materialism today is marked by the acceptance of *six* universal, developmental-logically consecutive *modes of production*.[41] In primitive societies, labor and distribution were organized by means of kinship relations. There was no private access to nature and to the means of production (primitive communal mode of production). In the early civilizations of Mesopotamia, Egypt, Ancient China, Ancient India, and pre-Columbian America, land was owned by the state and administered by the priesthood, the military, and the bureaucracy; this arrangement was superimposed upon the remains of village communal property (the so-called Asiatic mode of production). In Greece, Rome, and other Mediterranean societies, the private landowner combined the position of despotic master of slaves and day laborers in the framework of the household economy with that of a free citizen in the political community of city or state (ancient mode of production). In medieval Europe, feudalism was based on large private estates allotted to individual holders who stood in various relations of dependence (including serfdom) to the feudal lord; these relations were defined in terms that were at once political and economic (feudal mode of production). Finally, in capitalism, labor power became a commodity, so that the dependency of the immediate producers on the owners of the means of production was secured legally through the institution of the labor contract, and economically through the labor market.

The application of this schema runs into difficulties in anthropological and historical research. These are in part problems of mixed and transitional forms—there are only a few instances in which the economic structure of a specific society coincides with a single mode of production; both intercultural diffusion and temporal overlay permit complex structures to arise that have to be deciphered as a combination of several modes of production. But the more important problems are those posed by the developmental-logical ordering of the modes of production themselves.

If I am not mistaken, contemporary discussion revolves primarily around the following complexes:

a. It is not entirely clear how we can distinguish paleolithic from neolithic societies on the basis of the same primitive communal mode of production. The "neolithic revolution" signifies not only a new stage of development of productive forces but also a new mode of life.[42] For this reason, some have proposed distinguishing a stage of appropriative economy from a stage of producing economy. Whereas hunters and gatherers seized nature's treasures for their direct use, tillage and breeding already required means of production (earth and soil, livestock), which raised the question of ownership.[43] Other differences are related to the complexity of social organization (band, tribe, chiefdom).[44] Finally, it is possible to provide grounds for the conjecture that the technical innovations that marked the transition to neolithic society were dependent on the coherent development of mythological world views.[45]

b. The many-sided discussion of the so-called Asiatic mode of production has given rise to a whole series of systematic questions. Should this mode be understood as the last stage of the primitive communal order or as the first form of class society? [46] If the latter alternative can be made plausible—as I believe it can—does the Asiatic mode of production mark a universal stage of development or a special line of development of class societies *alongside of* the path of the ancient mode of production? Or is it a mixed form of the ancient and feudal modes of production? [47]

c. The classification of feudalism raises equally great difficulties.[48] Is this at all a clearly specifiable mode of production or merely a collective concept with no analytic pretensions? If there is an independent mode of production of this type, does it mark a universal stage of development? If so, did only the society of medieval Europe reach this stage; in other words, is feudalism a unique phenomenon, or did other civilizations also reach feudal stages of development?

d. This is connected with the further question, how can archaic civilizations be distinguished from developed civilizations? The differentiation of social subsystems and the increase in stratification took place within the framework of the same political class organization. In all evolutionarily successful civilizations there was a noteworthy structural change of world view—the change from a mythological-cosmogonic world view to a rationalized world view in the form of cosmological ethics. This change took place between the eighth and third centuries B.C. in China, India, Palestine, and Greece.[49] How can this be explained on materialist principles?

e. The controversy between theories of postindustrial society, on the one side, and theories of organized capitalism, on the other, also belongs in this context. It involves, among other things, the question of whether the capitalism regulated through state intervention in the developed industrial nations of the West marks the last phase of the old mode of production or the transition to a new one.

f. The classification of so-called socialist transitional societies is a special problem. Is bureaucratic socialism, compared to developed capitalism, in any sense an evolutionarily higher social formation; or are the two merely variants of the same stage of development?

These and similar problems have led as important a Marxist historian as Hobsbawm to cast doubt on the concept of *universal* stages of development (in his introduction to Marx's "Pre-Capitalist Economic Formations"). Of course, there remains the question of whether the aforementioned problems are merely lining the path of a normal scientific discussion or whether they are to be understood as signs of the unfruitfulness of a research program. I am of the opinion that the alternative should not be posed in this way at present. Perhaps the concept of a mode of production is not so much the wrong key to the logic of social development as a key that has not yet been sufficiently filed down.

V

The concept of a mode of production is not abstract enough to capture the universals of societal development. Modes of production can be compared at two levels: (a) regulation of access to the means of production, and (b) the structural compatibility of these rules with the stage of development of productive forces. On the first level, Marx differentiates according to whether property is communal or private. The viewpoint of exclusive disposition over the means of production leads, however, only to a demarcation of societies with and without class structures. Further differentiation according to the degree to which private property is established, and according to the forms of exploitation (the exploitation of village communities by the state, slavery, serfdom, wage labor), is as yet too imprecise to permit unambiguous comparisons.[50] To achieve greater precision, Finley recommends adopting the following points of view: claims to property versus power over things; power over human labor-force versus power over human movements; power to punish versus immunity from punishment; privileges and liabilities in judicial process; privileges in the area of the family; privileges of social mobility, horizontal and vertical; privileges versus duties in the sacral, political, and military spheres.[51] These general sociological points of view certainly permit a more concrete description of a given economic structure; but they broaden rather than deepen the analysis. The result of this procedure would be a pluralistic compartmentalization of modes of production and a weakening of their developmental logic. At the end of this inductivist path lies the surrender of the concept of the history of the species—and with it of historical materialism. The possibility that anthropological-historical research might one day force us to this cannot be excluded a priori. But in the meantime, the path leading in the opposite direction strikes me as not yet sufficiently explored.

It points in the direction of even stronger generalization, namely, the search for highly abstract principles of social organization. By principles of organization I understand innovations that become possible through developmental-logically reconstructible stages of learning, and which institutionalize new levels of societal learning.[52] The organizational principle of a society circumscribes ranges of possibility. It determines in particular: within which structures changes in the system of institutions are possible; to what extent the available capacities of productive forces are socially utilized and the development of new productive forces can be stimulated; to what extent system complexity and adaptive achievements can be heightened. A principle of organization consists of regulations so abstract that in the social formation which it determines a number of functionally equivalent modes of production are possible. Accordingly, the economic structure of a given society would have to be examined at two analytic levels: firstly in terms of the modes of production that have been concretely combined in it; and then in terms of that social formation to which the dominant mode of production belongs. A postulate of this sort is easier to put forward than to satisfy. I can only try to elucidate the research program and to make it plausible.

Organizational principles of society can be characterized, in a first approximation, through the institutional core that determines the dominant form of social integration. These institutional cores —kinship as a total institution, the state as a general political order, the complementary relation between a functionally specified state and a differentiated economic system—have not yet been thoroughly analyzed into their formal components. But I shall not follow this path of analysis here, since the formal components of these basic institutions lie in so many different dimensions that they can hardly be brought into a developmental-logical sequence. A more promising attempt can be made directly to classify, according to evolutionary features, the forms of social integration determined by principles of social organization.

Developmental-logical connections for the ontogenesis of action competence, particularly of moral consciousness, have already been rendered plausible. Of course, we ought not draw from ontogenesis over-hasty conclusions about the developmental levels of societies. It is the personality system that is the bearer of the ontogenetic learning process; and in a certain way, only social subjects can learn. But social systems, by drawing on the learning capacities of social subjects, can form new structures in order to solve steering problems that threaten their continued existence. To this extent the evolutionary learning process of societies is dependent on the competences of the individuals that belong to them. The latter in turn acquire their competences not as isolated monads but by growing into the symbolic structures of their life-worlds. This development passes through three stages of communication, which I would like to characterize now in a very rough way.

At the stage of *symbolically mediated interaction,* speaking and acting are still emeshed in the framework of a single, imperativist mode of communication. With the help of a communicative symbol, *A* expresses a behavioral expectation, to which *B* reacts with an action, in the intention of fulfilling *A*'s expectation. The meaning of the communicative symbol and of the action are reciprocally defined. The participants suppose that in interpersonal relations they could in principle exchange places; but they remain bound to their performative attitudes.

At the stage of *propositionally differentiated speech,* speaking and acting separate for the first time. *A* and *B* can connect the performative attitude of the participant with the propositional attitude of an observer; each can not only adopt the perspective of the other but can exchange the perspective of participant for that of observer. Thus two reciprocal behavioral expectations can be coordinated in such a way that they constitute a system of reciprocal motivation or, as we can also say, a social role. At this stage actions are separated from norms.

At the third stage, that of *argumentative speech,* the validity claims we connect with speech acts can be made thematic. In grounding assertions or justifying actions in discourse, we treat statements or norms (underlying the actions) hypothetically, that is, in such a way that they might or might not be the case, that they might be legitimate or illegitimate. Norms and roles appear as in need of justification; their validity can be contested or grounded with reference to principles.

I shall not deal with the cognitive aspects of this communicative development, but merely point out the step-by-step differentiation of a social reality graduated in itself. At first actions, motives (or behavioral expectations), and acting subjects are perceived on a single plane of reality. At the next stage actions and norms separate; norms draw together with actors and their motives on a plane that lies behind, so to speak, the reality plane of actions. At the last stage, principles with which norms of action can be generated are distinguished from these norms themselves; the principles, together with actors and their motives, are placed behind even the line of norms, that is, the existing system of action.

In this way we can obtain basic concepts for a genetic theory of action. These concepts can be read in two ways: either as concepts of the competences—acquired in stages—of speaking and acting subjects who grow into a symbolic universe or as concepts of the infrastructure of the action system itself. I would like to use them in this latter sense to characterize different forms of social integration. In doing so I shall distinguish the institutions that regulate the normal case from those special institutions, which, in cases of conflict, re-establish the endangered intersubjectivity of understanding (law and morality).

To the extent that action conflicts are not regulated through force or strategic means but on a consensual basis, there come into play structures that mark the moral consciousness of the individual and the legal and moral system of society. They comprise the core domain of the aforementioned general action structures—the representations of justice crystallizing around the reciprocity relation that underlies all interaction. In the Piagetian research tradition, developmental stages of moral consciousness have been uncovered which correspond to the stages of interactive competence.[53] At the *preconventional stage,* at which actions, motives, and acting subjects are still perceived on a single plane of reality, only the consequences of action are evaluated in cases of conflict. At the *conventional stage,* motives can be assessed independently of concrete action consequences; conformity with a certain social role or with an existing system of norms is the standard. At the *postconventional stage,* these systems of norms lose their quasi-natural validity; they require justification from universalistic points of view.

I have distinguished between general structures of action underlying the normal state (with little conflict) and those core structures that underlie the consensual regulation of conflicts. These structures of moral consciousness can find expression either in simply judging action conflicts or in actively resolving them. If at the same time we keep in mind the stages of development according to which these structures can be ordered, we can make intuitively plausible why there are often structural differences between these action domains; that is, (a) between the ability to master normal action situations and the ability to bring conflict situations under moral-legal points of view; and (b) between moral judgment and moral action. As in the behavior of the individual, stage differences also appear on the level of social systems. For example, in neolithic societies the moral and legal systems are at the preconventional stage of arbitration and feuding law; while normal situations (with little conflict) are regulated within the framework of the kinship system, that is, at the conventional stage. The situation is similar with structures of consciousness that are already clearly established in interpretive systems but have not yet found institutional embodiment in action systems. Thus in many myths of primitive societies there are already narratively constructed models of conflicts and their resolutions that correspond to the conventional stage of development of moral consciousness; at the same time the institutionalized law satisfies the criteria of the preconventional stage of moral consciousness.

In our (very tentative) attempt to distinguish *levels of social integration,* it is therefore advisable to keep separate (a) general structures of action, (b) structures of world views insofar as they are determinant for morality and law, and (c) structures of *institutionalized* law and of *binding* moral representations.

Neolithic Societies: (a) conventionally structured system of action (symbolic reality is graduated into the level of actions and that of norms); (b) mythological world views still immediately enmeshed with the system of action (with conventional patterns of resolving moral conflicts of action); (c) legal regulation of conflict from preconventional points of view (assessment of action consequences, compensation for resultant damages, restoration of status quo ante).

Early Civilizations: (a) conventionally structured system of action; (b) mythological world views, set off from the system of action, which take on legitimating functions for the occupants of positions of authority; (c) conflict regulation from the point of view of a conventional morality tied to the figure of the ruler who administers or represents justice (evaluation according to action intentions, transition from retaliation to punishment, from joint liability to individual liability).

Developed Civilizations: (a) conventionally structured system of action; (b) break with mythological thought, development of rationalized world views (with postconventional legal and moral representations); (c) conflict regulation from the point of view of a conventional morality detached from the reference person of the ruler (developed system of administering justice, tradition-dependent but systematized law).

The Modern Age: (a) postconventionally structured domains of action—differentiation of a universalistically regulated domain of strategic action (capitalist enterprise, bourgeois civil law), approaches to a political will-formation grounded in principles (formal democracy); (b) universalistically developed doctrines of legitimation (rational natural law); (c) conflict regulation from the point of view of a strict separation of legality and morality; general, formal, and rationalized law, private morality guided by principles.

VI

I would like now to illustrate how this approach can be made fruitful for the theory of social evolution. I shall choose the example of the emergence of class societies, since I can rely here on the aforementioned study by Klaus Eder.[54]

1. Class societies develop within the framework of a political order; social integration no longer needs to proceed through the kinship system; it can be taken over by the state. There have been a number of theories of the origin of the state, which I would like briefly to mention and to criticize.[55]

a. The *superimposition theory*[56] explains the emergence of a political ruling class and the establishment of a political order by nomadic tribes of herdsmen who subjugated sedentary farmers and set up a rule of conquerors. Today this theory is regarded as empirically refuted since nomadism appeared later than the first civilization.[57] The emergence of the state must have had endogenous causes.

b. The *division of labor theory*[58] is usually advanced in a complex version. Agricultural production achieved a surplus and led (in combination with demographic growth) to the freeing of labor forces. This made a social division of labor possible. The various social groups which thereby emerged appropriated social wealth differently and formed social classes, one (at least) of which assumed the functions of rule. Despite its suggestive power, this theory is not coherent. Social division of labor means functional specification within the vocational system; but vocational groups differentiated by knowledge and skill need not per se develop opposing interests that result in differential access to the means of production. There is no argument showing why functions of domination had to emerge from the constrast of interests rooted in vocational specialization. There was a social division of labor within the politically ruling class (the priesthood, military, and bureaucracy) as well as within the working population (e.g., between farmers and craftsmen).

c. The *inequality theory*[59] traces the emergence of the state directly to problems of distribution. With the productivity of labor there arose a surplus of goods and means of production. The growing differences in wealth resulted in social differences that a relatively egalitarian kinship system could not manage. The distribution problems required a different organization of social intercourse. If this thesis were correct, it could explain the emergence of system problems that could be solved by organization in a state; but this new form of social integration itself remains unexplained. Furthermore, the assumption of an automatic growth of productive forces is incorrect, at least for agricultural production. The Indians of the Amazon, for example, possessed all the technical means for producing a surplus in foodstuffs; but only contact with European settlers provided the impetus to use the available potential.[60] Among stock farmers there were, it is true, considerable inequalities, since herds can be enlarged rather easily.

d. The *irrigation hypothesis*[61] explains the merger of several village communities into a political unity by the desire to master the aridity of the land through large-scale irrigation systems. An administration was a functional requirement for the construction of such systems, and this administration became the institutional core of the state. This assumption has been empirically refuted since in Mesopotamia, China, and Mexico the formation of the state preceded the irrigation projects. Moreover, this theory would explain only the emergence of system problems and not the way in which they were resolved.

e. The *theory of population density*[62] explains the emergence of the state chiefly through ecological and demographic factors. One can assume an endogenous population growth that led normally to spatial expansion of segmentary societies, that is, to emigration to new areas. When, however, the ecological situation, adjoining mountains, the sea or the desert, barren tracts of land, or the like, hindered emigration or flight, conflicts were triggered by population density and the scarcity of land. This left no alternative but the subjugation of large segments of the population under the political domination of a victorious tribe. The complexity of densely populated settlements could be managed only through state organization. Even if population problems of this type could be demonstrated [to have existed] in *all* early civilizations, this theory, like the others, does not explain why and how these problems could be solved.

None of the above theories distinguishes between *system problems* that overload the adaptive capacity of the kinship system and the *evolutionary learning process* that explains the change to a new form of social integration. Only with the help of learning mechanisms can we explain why a few societies could find solutions to the steering problems that triggered their evolution, and why they could find precisely the solution of state organization. Thus I shall adopt the following orientations:

a. Developmental stages (in the sense of cognitive developmental psychology) can be distinguished in the ontogenesis of knowing and acting abilities. I understand these stages as learning levels that lay down the conditions for possible learning processes. Since the learning mechanisms belong to the equipment of the human organism (capable of speech), social evolution can rely on individual learning capacities only if the (in part phase-specific) boundary conditions are fulfilled.

b. The learning capacities first acquired by individual members of society or marginal social groups gain entrance into the interpretive system of the society through exemplary learning processes. Collectively shared structures of consciousness and stores of knowledge represent, in terms of empirical knowledge and moral-practical insight, a cognitive potential that can be used socially.

c. We may also speak of evolutionary learning processes on the part of societies insofar as they solve system problems that represent evolutionary challenges. These are problems that overload the adaptive capacities available within the limits of a given social formation. Societies can learn evolutionarily by utilizing the cognitive potential contained in world views for reorganizing action systems. This process can be represented as an institutional embodiment of rationality structures already developed in world views.

d. The introduction of a new principle of organization means the establishment of a new level of social integration. This in turn makes it possible to implement available (or to produce new) technical-organizational knowledge; it makes possible, that is, an increase in productive forces and an expansion of system complexity. Thus for social evolution, learning processes in the domain of moral-practical consciousness function as pacemakers.

2. With these as my points of orientation, I would like now to offer the following explanatory sketch of the origin of class societies.[63]

a. *The phenomenon to be explained* is the emergence of a political order that organized a society so that its members could belong to different lineages. The function of social integration passed from kinship relations to political relations. Collective identity was no longer represented in the figure of a common ancestor but in that of a common ruler.

b. *Theoretical explication of the phenomenon* A ruling position gave the right to exercise legitimate power. The legitimacy of power could not be based solely on authorization through kinship status; for claims based on family position, or on legitimate kinship relations in general, were limited precisely by the political power of the ruler. Legitimate power crystallized around the function of administering justice and around the position of the judge after the law was recognized in such a way that it possessed the characteristics of conventional morality. This was the case when the judge, instead of being bound as a mere referee to the contingent constellations of power of the involved parties, could judge according to intersubjectively recognized legal norms sanctified by tradition, when he took the intention of the agent into account as well as the concrete consequences of action, and when he was no longer guided by the ideas of reprisal for damages caused and restoration of a status quo ante, but punished the guilty party's violation of a rule. Legitimate power had in the first instance the form of a power to dispose of the means of sanction in a conventional administration of justice. At the same time, mythological world views also took on—in addition to their explantory function—justificatory functions, in the sense of legitimating domination.

c. *The goal of explanation follows from this* The differentiation of ruling positions presupposed that the presumptive ruler built legitimate power by virtue of a conventional administration of justice. Thus the emergence of the state should be explained through successful stabilization of a judicial position that permitted consensual regulation of action conflicts at the level of conventional morality.

The explanation sketch runs as follows:

d. *The initial state* I consider those neolithic societies in which the complexity of the kinship system had already led to a more strongly hierarchical organization to be the evolutionarily promising societies.

They had already institutionalized temporally limited political roles. The chieftains, kings, or leaders were judged by their concrete actions; their actions were not legitimate per se. Such roles were only temporarily institutionalized (e.g., for warfare) or limited to special tasks (e.g., to provide for rain and a good harvest). Viewed sociostructurally, these roles had not yet moved to the center of social organization.[64]

e. *Particular system problems*   In the evolutionarily promising neolithic societies system problems arose which could not be managed with an adaptive capacity limited by the kinship principle of organization. These might have been, for example, ecologically conditioned problems of land scarcity and population density or problems having to do with an unequal distribution of social wealth. These problems, irresolvable within the given framework, became more and more visible the more frequently they led to conflicts that overloaded the archaic legal institutions (courts of arbitration, feuding law).

f. *The testing of new structures*   A few societies under the pressure of evolutionary challenges from such problems made use of the cognitive potential in their world views and institutionalized—at first on a trial basis—an administration of justice at a conventional level. Thus, for example, the war chief was empowered to adjudicate cases of conflict, no longer only according to the concrete distribution of power, but according to socially recognized norms grounded in tradition. Law was no longer only that on which the parties could agree.

g. *Stabilization through the formation of systems*   These judicial positions could become the pacemakers of social evolution. However, as the example of the African Barotse empire shows, not all promising experiments led via such judicial functions to the permanent institutionalization of a ruling position, that is, to evolutionary success. Only under suitable conditions—such as, for example, the military victory of a tribe or construction of an irrigation project—could such roles be permanently differentiated, that is, stabilized in such a way that they became the core of a political subsystem. This marked off the evolutionarily successful from the merely promising social systems.

h. *The emergence of class structures*   "On the basis of political domination the material production process could then be uncoupled from the limiting conditions of the kinship system and reorganized via relations of domination." [65] The ruler secured the loyalty of his officials, of the priest and warrior families by assuring them privileged access to the means of production (palace and temple economy).

i. *Development of productive forces*   "The forces of production which were already discovered in the neolithic revolution could now be utilized on a large scale: the intensification of cultivation and stock-farming, and the expansion of the crafts were the results of the enlarged organizational capacity of class society. Thus there emerged new forms of cooperation (e.g., in irrigational farming) or of exchange (e.g., in the market exchange between town and country)." [66]

3. If it holds up empirically, this argument could also explain how opposing developments are connected in social evolution; namely, the cumulative learning process without which history could not be interpreted as evolution (i.e., as a directional process) and, on the other hand, the exploitation of man by man, which is intensified in class societies.[67] Historical materialism marked off linear progress along the axis of development of productive forces and adopted dialectical figures of thought for the development of productive relations. When we assume learning processes not only in the dimension of technically useful knowledge but also in that of moral-practical consciousness, we are maintaining [the existence of] developmental stages both for productive forces and for the forms of social integration. But the extent of exploitation and repression by no means stands in inverse proportion to these levels of development. Social integration accomplished via kinship relations and secured in cases of conflict by preconventional legal institutions belongs, from a developmental-logical point of view, to a lower stage than social integration accomplished via relations of domination and secured in cases of conflict by conventional legal institutions. Despite this progress, the exploitation and oppression *necessarily* practiced in political class societies has to be considered retrogressive in comparison with the less significant social inequalities *permitted* by the kinship system. Because of this, class societies are structurally unable to satisfy the need for legitimation that they themselves generate. This is, of course, the key to the social dynamic of class struggle. How is this *dialectic* of progress to be explained?

I see an explanation in the fact that new levels of learning mean not only expanded ranges of options but also new problem situations. A higher stage of development of productive forces and of social integration does bring relief from problems of the superseded social formation. But the problems that arise at the new stage of development can—insofar as they are at all comparable with the old ones—increase in intensity. This seems to be the case, at least intuitively, with the burdens that arise in the transition to societies organized through a state. On the other hand, the perspective from which we make this comparison is distorted so long as we do not also take into account the specific burdens of prestate societies; societies organized along kinship lines have to come off better if we examine them in the light of the kinds of problem first typical of class societies. The socialist battle-concepts of exploitation and oppression do not adequately discriminate among evolutionarily different problem situations. In [certain] heretical traditions one can indeed find suggestions for differentiating not only the concept of progress but that of exploitation. It is possible to differentiate according to bodily harm (hunger, exhaustion, illness), personal injury (degradation, servitude, fear), and finally spiritual desperation (loneliness, emptiness)—to which in turn there correspond various hopes—for well-being and security, freedom and dignity, happiness and fulfillment.

Excursus on Progress and Exploitation

I have tried to bring the basic institutions with which we can (to begin with) circumscribe principles of social organization—family, state, differentiated economic system—into relation with historical progress via developmental stages of social integration. But evolutionarily important innovations mean not only a new level of learning but a new problem situation as well, that is, a new category of burdens that accompany the new social formation. The dialectic of progress can be seen in the fact that with the acquisition of problemsolving abilities new problem situations come to consciousness. For instance, as natural-scientific medicine brings a few diseases under control, there arises a consciousness of contingency in relation to all illness. This reflexive experience is captured in the concept of *quasi-nature* [*Naturwüchsigkeit*]—an area of life having been seen through in its pseudo-naturalness is quasi-natural. Suffering from the contingencies of an uncontrolled process gains a new quality to the extent that we believe ourselves capable of rationally intervening in it. This suffering is then the negative of a new need. Thus we can make an attempt to interpret social evolution taking as our guide those problems and needs that are first brought about by evolutionary advances. At every stage of development the social-evolutionary learning process itself generates new resources, which mean new dimensions of scarcity and thus new historical needs.

With the transition to the sociocultural form of life, that is, with the introduction of the family structure, there arose *the problem of demarcating society from external nature.* In neolithic societies, at the latest, harmonizing society with the natural environment became thematic. Power over nature came into consciousness as a scarce resource. The experience of powerlessness in relation to the contingencies of external nature had to be interpreted away in myth and magic. With the introduction of a collective political order, there arose *the problem of the self-regulation of the social system.* In developed civilizations, at the latest, the achievement of order by the state became a central need. Legal security came to consciousness as a scarce resource. The experience of social repression and arbitrariness had to be balanced with legitimations of domination. This was accomplished in the framework of rationalized world views (through which, moreover, the central problem of the previous stage—powerlessness—could be defused). In the modern age, with the autonomization of the economy (and complementarization of the state), there arose *the problem of a self-regulated exchange of the social system with external nature.* In industrial capitalism, at the latest, society consciously placed itself under the imperatives of economic growth and increasing wealth. Value came into consciousness as a scarce resource. The experience of social inequality called into being social movements and corresponding strategies of appeasement. These seemed to lead to their goal in social welfare state mass democracies (in which, moreover, the central problem of the preceding stage—legal insecurity—could be defused). Finally, if postmodern societies, as they are today envisioned from different angles, should be characterized by a primacy of the scientific and educational systems, one can speculate about the emergence of *the problem of a self-regulated exchange of society with internal nature.* Again a scarce resource would become thematic—not the supply of power, security, or value, but the supply of motivation and meaning. To the extent that the social integration of internal nature—the previously quasi-natural process of interpreting needs—was accomplished discursively, principles of participation could come to the fore in many areas of social life; whereas the simultaneously increasing dangers of *anomie* (and *acedie*) could call forth new administrations concerned with motivational control. Perhaps a new institutional core would then take shape around a new organizational principle, an institutional core in which there merge elements of public education, social welfare, liberalized punishment, and therapy for mental illness.

I mention this perspective—for which there exist clues at best —only to elucidate the *possibility* that a sociostructurally anchored pattern of differential exercise of social power could outlive even the *economic form* of class domination (whether exercised through private property rights or state bureaucracies occupied by elites). In a future form of class domination, softened and at the same time intensified, to sociopsychological coercion, "domination" (*Herrschaft*—the term calls to mind the open, person-bound, political form of exercising social force, especially that of European feudalism) would be refracted for a second time, not through bourgeois civil war, but through the educational system of the social welfare state. Whether this would *necessarily* give rise to a vicious circle between expanded participation and increasing social administration, between the process of motive formation becoming reflective and the increase in social control (i.e., in the manipulation of motives) is, in my opinion, a question that cannot be decided in advance (despite the confident judgment of revivified pessimistic anthropologies).

I have proposed a spectrum of problems connected with the self-constitution of society, ranging from demarcation in relation to the environment, through self-regulation and self-regulated exchange with external nature, to self-regulated exchange with internal nature. With each evolutionarily new problem situation there arise new scarcities, scarcities of technically feasible power, politically established security, economically produced value, and culturally supplied meaning; and thus new historical needs come to the fore. If this bold schema is plausible, it follows that the logical space for evolutionarily new problem domains is exhausted with the reflexive turn of motive formation and the structural scarcity of meaning; the end of the *first* run-through could mean a return, at a new level, to problems of demarcation—namely, to the discovery of internal limits which the socialization process runs up against—and to *the outbreak of new contingencies at these limits of social individuation.*

VII

In closing, I would like to indicate the perspectives that arise for dealing with competing explanatory approaches. Structuralism, neoevolutionism, and sociological functionalism have been put forward as approaches to evolution theory. In addition, the concept of historical progress, which is closely connected with that of social evolution, raises questions relating to the logic of science; these questions have been dealt with in the form of a critique of the philosophy of history[68] and, on the other hand, in the framework of an evolutionary ethics.[69]

1. Althusser and Godelier have tried to bring the concepts and assumptions developed by Levi-Strauss into historical material-ism.[70] The concept of structure was developed in dealing with primitive societies, in connection with both the analogical structures of "the savage mind" and the familial structures of social relations. The concept refers to basic systems of rules that are followed in cognition, speech, and interaction. These rules cannot be directly read off the surface of phenomena; they are rather deep structures, which individuals follow nonintentionally in generating observable cultural formations. The rules are not only valid for single individuals; they have collective validity as well. Moreover, in each case they form a system that makes it possible to establish transformation relations between the expressions generated. The structures can be rationally reconstructed.[71]

I cannot now go into the various attempts that have been made to adopt basic structuralist concepts for Marxist purposes. They have promoted an inflationary employment of these concepts beyond the well-circumscribed domain of anthropology; thus clear definitions are called for. On the level of the personality system we can delimit three structural dimensions from one another: cognition, speech, and interaction. This means that the individual develops structures and corresponding competences which make possible (a) operations of thought, of cognitively processing experiences, and of instrumental action; (b) the production of phonetically and grammatically well-formed sentences; and (c) interactions, as well as consensual regulation of action conflicts. On the other hand, communication in language (and, in a different way, strategic action as well) requires an integration of structures from more than one of these dimensions. For this reason, the structures of *utterances* in language—going beyond the [narrowly] linguistic—are not easy to analyze. The significance of the medium of language is evident; in it individual and social consciousness are combined.

On the level of the social system we can, if I am not mistaken, specify distinctive elementary deep structures for productive forces and for the forms of social integration. Forces of production incorporate technical and organizational knowledge, which can be analyzed in terms of cognitive structures. The institutional framework and the mechanisms for conflict regulation incorporate practical knowledge, which can be analyzed in terms of structures of interaction and forms of moral consciousness. World views, by contrast, are highly complex formations that are determined by cognitive, linguistic, and moral-practical forms of consciousness; the composition and the interplay of the structures is not fixed once and for all.

The attempts at rational reconstruction have hitherto flourished primarily in areas in which elementary deep structures are easier to isolate: in linguistics, that is, in phonetic and syntactic theory; in anthropology, insofar as it is concerned with primitive kinship systems (mythological world views are accessible to structuralist analysis to the degree that they are still enmeshed with structures of interaction); [72] and finally, it is also fruitful in psychology, insofar as it is concerned (in the Piagetian tradition of research) with the ontogenesis of thought and of moral consciousness.[73] Attempts at reconstruction have been less successful in areas in which several structures work together: this can be seen in pragmatic theory; in sociolinguistics and ethnolinguistics, insofar as they are concerned with universals of processes of uttering and understanding; in the psychoanalytic theory of language, which investigates the conditions of systematically distorted communication; and, finally, in the structuralist analysis of world views, which seldom penetrates beyond the surface of complex traditions.[74]

Structuralism has naturally come up against the limits of all synchronic investigations. In linguistics and anthropology this has been less noticeable only because of the static properties of their object domains. For the most part, structuralism limits itself to the logic of existing structures and does not extend to the pattern of structure-forming processes. Only the genetic structuralism worked out by Piaget, which investigates the developmental logic behind the process in which structures are formed, builds a bridge to historical materialism. As shown above, it offers the possibility of bringing different modes of production under abstract developmental-logical viewpoints.

It is indeed possible to model the history of technology on the ontogenetically analyzed stages of cognitive development, so that the logic of the development of productive forces becomes visible. But the historical sequence of modes of production can be analyzed in terms of abstract principles of social organization only if we can specify which structures of world views correspond to individual forms of social integration and how these structures limit the development of secular knowledge. In other words, precisely a historical-materialist appproach is directed to a structural analysis of the development of world views. The evolution of world views mediates between the stages of development of interaction structures and advances in technically useful knowledge. In the concepts of historical materialism this means that the dialectic of forces and relations of production takes place through ideologies.

2. The anthropological theories of evolution of the late nineteenth century (Morgan, Tylor) were driven back, in our century, by the culture-relativistic views of the functionalist school; only authors like V. G. Childe and L. White held on to the concept of general stages of development.[75] Under the influence of the

dominant cultural anthropology (Kroeber, Malinowski, Mead), developmental-theoretic views were—as the "multilinear evolutionism" of a J. H. Steward shows[76]—represented only in a very cautious form and accommodated to cultural ecology. More recently, however, the success of the theory of biological evolution has again given impetus to the renewal of social-scientific evolutionism. Social evolution no longer appears only vaguely as a continuation of organic evolution; instead neoevolutionists (Parsons, Luhmann, Lenski)[77] start with the idea that social evolution can be explained in accord with the well-analyzed and well-tested model of natural evolution. The heuristic usefulness of the biological model is not at issue; it is however doubtful whether it points the way to a generalized theory of evolution valid for both natural and cultural development.[78]

As we know, the biological model relies on the concept of the self-maintenance of self-regulating systems that demarcate themselves from hypercomplex environments. Between the environment and the system there is a complexity gap; the boundary-maintaining system is faced with the task of developing as much self-complexity as is needed to enable it adequately to reduce the complexity of the environment. The bearers of natural evolution are the species, each of which is represented by a specific genetic makeup capable of reproducing itself. Species reproduce themselves in the form of populations that stabilize themselves in their ecological surroundings. These in turn are composed of individual organisms that interact among themselves and with the environment. The evolutionary learning process applies immediately to the genetic makeup. Through the process of mutation, which can be understood as an error in the transmission of genetic information, divergent phenotypes are produced; under the selection pressure of the environment these are selected for, making possible the stabilization of a population dependent on the conditions in its environment. This nonteleologically steered learning process leads to a result that can be teleologically interpreted— the species can be rank-ordered from morphological and behavioristic viewpoints, that is, according to the complexity of their physical organization and the range of their reaction potential.

In carrying this model over to social development, three basic

problems arise: What is the equivalent for the process of mutation? What is the equivalent for the ability of a population to
survive? Finally, what is the equivalent for the evolutionary step-
ladder occupied by the various species?

a. In my view, the heuristic usefulness of the biological model
consists in its directing our attention to the evolutionary learning
mechanism. At the basis of cultural tradition there is evidently
a variety-generating mechanism that, in what is for the time being
a vague sense, corresponds to mutation. Natural evolution is not
affected by those individual learning processes of individual organisms that extend and modify genetically programmed behavior (for the behavioral modification is limited to the life-cycle
of the individual organism and not fed back into the next round
of reproduction of the genetic makeup). By contrast, at the sociocultural stage of development learning processes are socially organized from the start, so that the results of learning can be
handed down. Thus cultural tradition provides a medium through
which variety-generating innovations can operate after the mechanism of natural evolution has come to rest.

The differences between mutation processes and social learning
leap to the eye.[79] In the case of social evolution the learning
process takes place not through changes in genetic makeup but
through changes in knowledge potential. The distinction between
phenotype and genotype loses its meaning at this level. The
intersubjectively shared knowledge that is passed on is part of
the social system and not the property of isolated individuals; for
they become individuals only in the process of socialization. Natural evolution leads to a more or less homogeneous repertoire of
behavior among the members of a species, whereas social learning
results in an accelerated diversification of behavior. These comparisons could be continued. But I see a fundamental difficulty
in the fact that while biochemistry has recently met with success
in analyzing the process of mutation, the learning mechanism
that is at the basis of so complex a phenomenon as cultural tradition is almost unknown. Again cognitive and analytical developmental psychology hold out some promise here; as learning
mechanisms they propose either accommodation and assimilation

in the learning of new cognitive structures or identification and projection in the construction of a motivational basis. As long as these mechanisms are not adequately analyzed, however, we cannot judge whether the comparison between mutation and tradition is merely metaphorical or whether the underlying social learning mechanism is in some ways functionally equivalent to the process of mutation. One difference should arouse our suspicion; whereas the mutation process produces chance variations, the ontogenesis of structures of consciousness is a highly selective and directional process.

b. In natural evolution the success of learning processes is measured against the ability of a population to stabilize itself in a given environment; and the reproduction of the species depends, in the final analysis, on the individual organism's ability to survive. We can specify in turn unambiguous parameters for the ability of an organism to avoid death. This is not the case for the ability of a society to avoid death; it is not even clear what this is supposed to mean. The physical survival of a number of members sufficient for reproduction is, of course, a necessary condition for a society's maintaining its identity—but it is not its sufficient condition.

The identity of a society is normatively determined and depends on cultural values; on the other hand, these values can change as the result of a learning process. There is no clearly specifiable goal-function against which the ultrastability of societies could be measured. Dunn gives the following formulation to this state of affairs:

The appropriateness of novel behavior is tested by its contribution to goal convergence. If it fails that test it will usually fail to win a permanent place in the behavioral repertoire. However, the failure to generate goal convergence may not only cause the new behavioral mode to be identified as maladaptive, it may also call into question the appropriateness of the goal. In short, just as the goals form the test of adaptive behavior giving rise to the revision of behavioral ideas, behavioral ideas sometimes form a test of the adequacy of goals and lead to goal revision.[80]

I shall not pursue the proposals of Dunn and Luhmann[81] for an evolutionary assessment of the highest system values (system target goals) because they do not lead us out of the hopeless circle of a self-referential definition of social life. At the sociocultural stage, learning processes are from the outset linguistically organized, so that the objectivity of the individual's experience is structurally entwined with the intersubjectivity of understanding among individuals. For this reason the relation between socialized individuals and their society is not the same instrumental relation as that between exemplar and species at infrahuman stages of development. It is also senseless to propose instrumentalizing the highest system values with a view to what the individuals in question know and want; for these individuals have been socialized in their society. If there should be normative viewpoints for the ultrastability of societies, we might at most seek them in those basic structures of linguistic communication in which societies reproduce themselves together with their members. Species reproduce themselves when sufficiently many exemplars avoid death; societies reproduce themselves when they avoid passing on too many errors. If the survival ability of organisms is a test case for the learning process of the species, then the corresponding test cases for societies lie in the dimension of the production and utilization of technically and practically useful knowledge.

c. Finally, in carrying over the biological model to historical development, there is also a difficulty in the fact that the viewpoint of increasing complexity does not suffice for making out evolutionary thresholds or levels of development. Dunn proposes distinguishing three stages of social development: in the first stage, the social system expends its entire adaptive capacity in dealing with the risks of external nature; in the second stage, more adaptive achievements are required for dealing with other social systems than for mastering nature; in the third stage, the adaptive achievements that were developed in dealing with the natural and social environments become reflective: the learning of learning.[82] Luhmann proposes that the division be undertaken according to the degree of differentiation of the three basic evolutionary functions, that is, according to the step-by-step separation of variation, selection, and stabilization. Even if these criteria could be applied to the historical material, they are unsatisfactory; from functionalist points of view we can indeed distinguish *degrees* of complexity, but not *stages* of evolution.

Even in natural evolution the degree of complexity is not a sufficient condition for placing a species in the evolutionary rank order; for increasing complexity in physical organization or mode of life often proves to be an evolutionary dead end. A reliable evolutionary classification is possible only if we know the inner logic of a series of morphological changes or of an expansion of reaction potential. The role played by the central nervous system in phylogenetic comparison is prototypical here; we have to know the general structure and logic of development of the CNS if we want to classify different species according to the state of development of this system.[83] In social evolution as well, we shall not be able to classify social formations according to their state of development until we know the general structures and developmental logic of social learning processes. Corresponding to the central nervous system here are the basic cognitive structures in which technical and moral-practical knowledge are produced.

3. Social-scientific neoevolutionism is usually satisfied with the directional criterion of increasing steering (or adaptive) capacity. From this vantage point, the concepts and problems of a functionalism developed along systems-theoretic lines are brought into developmental theory. Modernization theories, for example, move within this methodological framework. The combination of the conceptual repertoires of systems theory and evolution theory is undoubtedly advantageous in investigating structural changes that expand the steering capacity of a society. On the other hand, this analytic gain has misled [some] to confuse structures of learning capability with social complexity. A self-sufficient functionalism fails to appreciate the fact that increases in complexity are in each case possible only at the learning level attained in the organizational principle of the society in question. But we cannot explain the establishment of new organizational principles without knowing the basic structures specific to processes of socializa-

tion, as well as the logic of development of these structures. An elucidation of the learning abilities specific to object domains has to precede the analysis of complexity.

This can be seen, for example, in the use of the systems-theoretic concept of communication media. The fundamental medium is evidently language. The fixation of speech in writing was an evolutionarily significant step. Another was the differentiation of subsystems established through special media: the political system through law, the economic system through money, the scientific system through truth, and so on.[84] Functional analysis can only show here *that* such innovations increase the complexity of society; it does not explain *how* the development of communication media on the basis of language is *structurally possible;* just as little does it explain *why* specific media are introduced in a given form of social integration. I cannot so much as indicate here how a theory of communication might derive the various media from basic structures of speaking and acting; but I would like at least to point out one consequence.

Only if we succeed in ordering a series of organizational principles according to a developmental logic and in specifying corresponding stages of social evolution, can the analysis of complexity find its proper place. It would then serve to explain the *special evolution* that societies undergo in adapting to ecological conditions and historical circumstances. If we could not supplement genetic-structuralist research into general evolution with a functionalistically oriented examination of special evolutions, the sociocultural morphology of individual societies would necessarily escape the theory of evolution.[85]

4. At the conclusion of our reflections I would like to return again to the normative implications which every theory of development has; even the theory of natural evolution has to provide a directional criterion that makes it possible to assess morphological properties and reaction capabilities. The choice of this criterion appears to be less problematic in the case of natural evolution only because we can fall back on the basic value of "survival" (or "health"). Organic life is so synonymous with the reproduction of this life that we attribute the normative distinc-

tion of all healthy states not to the observer but to the living systems themselves. In living, the organisms themselves make an evaluation to the effect that self-maintenance is preferable to the destruction of the system, reproduction of life to death, health to the risks of sickness. The theorist of evolution feels himself relieved of value judgments; he seems to be merely repeating the "value judgment" that is given with the form of reproduction of organic life. This of, course, a logical error; from the descriptive statement that living systems prefer certain states to others there in no way follows a positive evaluation by the observer.

Can one say perhaps that the theorist of evolution, because he is himself a living being, is spontaneously inclined not merely to observe the normative distinction of the avoidance of death as a natural phenomena, but also to agree with it? In any case, only this agreement justifies the attitude of many biologists, who regard the direction of evolution as something good, and not only distinguish but evaluate the species according to the place they hold in the evolutionary rank order. Only under this presupposition, at any rate, are the attempts to develop an evolutionary ethics comprehensible.[86]

In C. H. Waddington's version, evolutionary ethics is based on the metaethical insight of the biologist ("biological wisdom") "that the function of ethical beliefs is to mediate human evolution, and that evolution exhibits some recognizable direction of progress." [87] Waddington believes he can avoid a naturalistic fallacy:

I argue that if we investigate by normal scientific methods the way in which the existence of ethical beliefs is involved in the causal nexus of the world's happenings, we shall be forced to conclude that the function of ethicizing is to mediate the progress of human evolution, a progress which now takes place mainly in the social and psychological sphere. We shall also find that this progress, in the world as a whole, exhibits a direction which is as well or ill defined as the concept of physiological health. Putting these two points together we can define a criterion, which does not depend for its validity on any recognition by a preexisting ethical belief.[88]

But if the biological wisdom of any ethics singled out by evolution is expressed in the fact that it promotes the evolution and the learning ability of social systems, then we have to presuppose

(a) that we know how social evolution can be measured, and (b) that we regard social evolution as good. Waddington starts from the idea that these presuppositions have been adequately clarified within biology because (a) the directional criterion of natural evolution is supposed to hold for social evolution as well, and (b) with the reproduction of life, health is posited as an objective value. Even if (a) were unproblematic, there is in (b) a naturalistic fallacy: the biologist is in no way forced to adopt as his own preference the observed tendency to self-maintenance inherent in organic life—unless it be through the fact that he is himself a living being. But in the objectivating attitude of the knowing subject he can ignore this fact.

The situation is somewhat different in the case of the normative foundation of linguistic communication, upon which, as theoreticians, we must always (already) rely. In adopting a theoretical attitude, in engaging in discourse—or for that matter in any communicative action whatsoever— we have always (already) made, at least implicitly, certain presuppositions, under which alone consensus is possible: the presupposition, for instance, that true propositions are preferable to false ones, and that right (i.e., justifiable) norms are preferable to wrong ones. For a living being that maintains itself in the structures of ordinary language communication, the validity basis of speech has the binding force of universal and unavoidable—in this sense transcendental—presuppositions.[89] The *theoretician* does not have the same possibility of choice in relation to the validity claims immanent in speech as he does in relation to the basic biological value of health. Otherwise he would have to deny the very presuppositions without which the theory of evolution would be meaningless. If we are not free then to reject or to accept the validity claims bound up with the cognitive potential of the human species, it is senseless to want to "decide" for or against reason, for or against the expansion of the potential of reasoned action.[90] For these reasons I do not regard the choice of the historical-materialist criterion of progress as arbitrary. The development of productive forces, in conjunction with the maturity of the forms of social integration, means progress of learning ability in both dimensions: progress in objectivating knowledge and in moral-practical insight.

5 Legitimation Problems in the Modern State

This paper was presented at a meeting of the *Deutsche Vereinigung für Politische Wissenschraft* in October of 1974. Remarks referring to or based on the occasion have been omitted.

To know whereof one speaks is always beneficial; this is especially true when dealing with the problem of legitimacy. . . . After (1) making a few conceptual distinctions, I would like (2) to examine the principle of legitimacy in modern times. I shall then show (3) how the modern problem of legitimacy arises from structures of the bourgeois state and (4) how the problem shifts in developed capitalist states. In the final section I shall (5) examine several concepts of legitimation with the aim of justifying the reconstructive concept used here.

I

Legitimacy means that there are good arguments for a political order's claim to be recognized as right and just; a legitimate order deserves recognition. *Legitimacy means a political order's worthiness to be recognized.* This definition highlights the fact that legitimacy is a contestable validity claim; the stability of the order of domination (also) depends on its (at least) de facto recognition. Thus, historically as well as analytically, the concept is used above all in situations in which the legitimacy of an order is disputed, in which, as we say, legitimation problems arise. One side denies, the other asserts legitimacy. This is a *process*—Talleyrand endeavored to *legitimize* the House of Bourbon. Processes of this kind were rendered less dramatic in the modern constitutional state (with the institutionalization of an opposition); that is, they were defused and normalized. For this reason it is realistic to speak today of legitimation as a permanent problem. Of course, in this framework too, legitimation conflicts flare up only over questions of principle (as, for example, in 1864 over the budgetary rights of the Prussian Landtag). Such conflicts can lead to a temporary withdrawal of legitimation; and this can in certain circumstances have consequences that threaten the continued existence of a regime. If the outcome of such legitimation crises is connected with a change of the basic institutions not only of the state but of the society as whole, we speak of revolutions. (It does not serve to clairfy matters when the Reformation or the introduction of the mechanical loom or German Idealism are called revolutions, thus inflating the term.)

Less trivial is the *domain of application* of the concept of legitimacy. Only political orders can have and lose legitimacy; only they need legitimation. Multinational corporations or the world market are not capable of legitimation. This is true also of prestate, so-called primitive, societies that are organized according to kinship relations. To be sure, in these societies there are myths that interpret the natural and social order. They fix membership in the tribal group (and its limits) and thus secure a collective identity. Mythological world views here have a constitutive significance rather than a subsequent legitimating significance.[1]

We first speak of legitimacy in relation to political orders. Historically political domination crystallized around the function of the royal judge, around the nucleus of conflict regulation on the basis of recognized legal norms (and no longer only through the force of arbitration). The administration of justice at this level establishes a position that owes its authority to disposition over a legal system's force of sanction and no longer only to kinship status (and to the mediator role of the arbitrator). The legitimate power of the judge can become the nucleus of a system of domination to which the society gives over the function of intervening when its integrity is threatened.[2] The state does not, it is true, itself establish the collective identity of the society; nor can it itself carry out social integration through values and norms, which are not at its disposition. But inasmuch as the state assumes the guarantee to prevent social disintegration by way of binding decisions, the exercise of state power is tied to the claim of maintaining society in its normatively determined identity. The legitimacy of state power is then measured against this; and it must be recognized as legitimate if it is to last.

In more recent theories of political development, which attempt to explain the emergence of the modern state, securing identity, procuring legitimation, and social integration are listed as general system problems.[3] Of course, the systems-theoretic reformulation of these concepts conceals the connection that is constitutive for political domination. The political subsystem takes on the task of protecting society from disintegration; but it cannot freely dispose of the capacities of social integration or of the definitional power through which the identity of the society is fixed. At the evolutionary stage of societies organized through a state, different forms of identity have developed: the empire, the city state, the nation state. These are, to be sure, only compatible with certain types of political domination, but they do not coincide with them. A world empire, a polis, a medieval commune, a nation—these express the connection of different political orders with different forms of life (ethos).[4] Thus modernization research is correct in taking state-building and nation-building as two different, if interdependent, processes.

The restriction of the category of legitimacy to societies organized through a state is not trivial. This conceptual specification has empirical implications; I would like to mention the following points.

a. If we equate legitimate power with political domination, we have to maintain, among other things, that no political system can succeed in permanently securing mass loyalty—that is, its members' willingness to follow—without recourse to legitimations. In the many-sided discussion of Max Weber's type of legal domination, which is supposed to legitimize itself solely through technical procedures, only Carl Schmitt and Niklas Luhmann come close to the position that in the modern state decisions legally arrived at are accepted, so to speak, without motives. On a somewhat different level we find the position that the social integration achieved through values and norms and protected by the authority of the state could in principle be replaced by system integration, that is, by the latent functions of nonnormative social structures (or mechanisms).[5] Corresponding to this is the assertion that system performance can render representations of legitimacy superfluous, that the neutrally *observable* efficiency of the state apparatus or of the economic system (and not only the efficiency perceived and evaluated by participants) is effective for legitimation.[6] These assertions are incompatible with the proposed usage of the concept of legitimacy.

b. Furthermore, according to this usage, problems of legitimacy are not a specialty of modern times. The formulas of *legitimum imperium* or *legitimum dominium* were widespread in Rome and in the European Middle Ages.[7] Political theories occupied themselves with the issue of the rise and fall of legitimate domination, in Europe at the latest since Aristotle, if not since Solon.[8] And we can demonstrate the existence of legitimacy conflicts themselves in all older civilizations, even in archaic societies, when, in the wake of colonization, they collide with conquerors from societies organized through states. In traditional societies, legitimation conflicts typically take the form of prophetic and messianic movements that turn against the official version of religious doctrine, which legitimates the state or a priestly domination, the church or a colonial domination. In the process the insurgents appeal to the original religious content of the doctrine— examples would be the prophetic movements in Israel, the spread of primitive Christianity in the Roman Empire, the heretical movements of the Middle Ages up to the Peasants' War, but also the messianic, millenarian movements among indigenous populations who took the religion of their colonial masters only to turn it against them, criticizing their legitimacy. V. Lanternari cites the revealing saying of a Zulu prophet: "At first we had the land and you the Bible; now you have the land and we are left with the Bible."[9] I cannot understand how, in the face of these world-wide phenomena, one could insist on reserving legitimation problems to bourgeois society and the modern state.

c. I find even less comprehensible the assertion that legitimation problems have nothing to do with class conflicts. With the differentiation of a political control center, there arose the possibility of uncoupling access to the means of production and appropriation of socially produced wealth from the kinship system and reorganizing them in accord with relations of domination. And this structural possibility was also made use of in all civilizations. A class structure thereby emerged, which of course did not yet appear *as* a socioeconomic division of classes but rather as a structure of privilege of estates, castes, ranks, and so on. From all indications, stratification, exploitation, and face-to-face social force had reached an advanced state in the old empires. One has only to study the history of penal law to see that structural conflicts were built into these traditional societies, conflicts that had to break out in crises again and again. In this connection one might simply take a look at Rostovtzeff's chapter on the Gracchi and the beginnings of political and social upheavals in Rome.[10]

In the European Middle Ages revolts by peasants, journeymen, and urban communities were widespead; many of them did not go beyond the thresholds critical for legitimation; but they often did when they were connected with heretical movements. Examples would be the Brothers and Sisters of the Holy Ghost, a pantheistic sect that developed around 1300 on both sides of the lower Rhine,[11] or the radical Franciscans in the northern Italian cities of the fourteenth century.[12] The Peasants' War was only the last important link in a long chain of heretically founded and socially motivated movements.[13] There is no need to waste words on the class background of the bourgeois revolutions.

It is not surprising that class confrontations lay behind the different manifestations of delegitimation; for state organization of society is the most important condition for a class structure in the Marxian sense. Naturally legitimacy conflicts were not as a rule carried out in terms of economic conflicts, but on the level of the legitimating doctrines. They had to relate to definitions of collective identity; and these could in turn be based only on structures that established unity and guaranteed consensus, like language, ethnic background, tradition, or, indeed, reason. (The only exception I know of is the Communist Party, which determined for a time the identity of the labor movement. But even it is a structure that produces dissension only in the first instance; the goal of the movement led by the Communist Party is supposed to involve making itself superfluous as a party.)

Let me now briefly summarize the results of our conceptual analysis. By *legitimacy* I understand the worthiness of a political order to be recognized. The *claim to legitimacy* is related to the social-integrative preservation of a normatively determined social identity. *Legitimations* serve to make good this claim, that is, to show how and why existing (or recommended) institutions are fit to employ political power in such a way that the values constitutive for the identity of the society will be realized. Whether legitimations are convincing, whether they are believed, depends naturally on empirical motives; but these motives are not formed independently of the (formally analyzable) justificatory force of the legitimations themselves. We can also say that they are not independent of the legitimation potential, of the *grounds or reasons,* that can be mobilized. What are accepted as reasons and have the power to produce consensus, and thereby to shape motives, depends on the *level of justification* required in a given situation. Since I would like to use the concept of legitimation in a reconstructive manner, I shall take up briefly the question of the internal structure of justifications.

II

P. von Kielmannsegg has provided clear criticisms of Weberian types of legitimacy and proposed that we understand traditionalism and charisma as states that *every* legitimate order can assume. We can distinguish these aspects of the establishment and maintenance of legitimate power from the forms of legitimate power, the types of domination. Here again we can separate the *legitimating grounds* from the *institutionalizations of domination.* Certain systems of institutions are compatible with a given level of justification; others are not.

I cannot characterize the historically familiar levels of justification in terms of their formal properties (as would be necessary); instead I shall illustrate them with a few allusions. In early civilizations the ruling families justified themselves with the help of myths of origin. Thus the pharoahs represented themselves first as gods—for example, as the god Horus, son of Osiris. On this level narrative grounds are sufficient, viz. mythological stories. With the imperial development of the ancient civilizations the need for legitimation grew; now not only the person of the ruler had to be justified, but a political order (against which the ruler could transgress). This end was served by cosmologically grounded ethics, higher religions, and philosophies, which go back to the great founders: Confucius, Buddha, Socrates, the prophets of Israel, and Jesus.[14] These rationalized world views had the form of dogmatizable knowledge. Arguments took the place of narratives. There were to be sure ultimate grounds, unifying principles, which explained the world as a whole (the natural and human world). The ontological tradition of thought was also on this level. Finally, in modern times, especially since the rise of modern science, we learned to distinguish more strictly between theoretical and practical argumentation. The status of ultimate grounds became problematic. Classical natural law was reconstructed; the new theories of natural law that legitimated the emerging modern state claimed to be valid independently of cosmologies, religions, or ontologies.

With Rousseau and Kant this development led to the conclusion that the formal principle of reason replaced material principles like Nature or God in practical questions, questions concerning the justification of norms and actions. Here justifications are not based only on arguments—that was also the case in the framework of philosophically formed world views. Since ultimate grounds can no longer be made plausible, *the formal conditions of justification themselves obtain legitimating force.* The procedures and presuppositions of rational agreement themselves become principles. In contract theories, from Hobbes and Locke to John Rawls,[15] the fiction of a state of nature or of an original position *also* has the meaning of specifying the conditions under which an agreement will express the common interest of all involved—and to this extent can count as rational. In transcendentally oriented theories, from Kant to Karl-Otto Apel,[16] these conditions, as universal and unavoidable presuppositions of rational will-formation, are transposed either into the subject as such or into the ideal communication community. In both traditions, it is the formal conditions of possible consensus formation, rather than ultimate grounds, which possess legitimating force.

Thus, by *levels of justification* I mean formal conditions for the acceptability of grounds or reasons, conditions that lend to legitimations their efficacy, their power to produce consensus and shape motives. These levels can be ordered hierarchically. The legitimations of a superseded stage, no matter what their content, are depreciated with the transition to the next higher stage; it is not this or that reason which is no longer convincing but the *kind* of reason. Such depreciation of the legitimation potential of entire blocks of tradition occurred in civilizations with the retrenchment of mythological thought, and in modern times with the retrenchment of cosmological, religious, and ontological modes of thought. My conjecture is that these depreciatory shifts are connected with social-evolutionary transitions to new learning levels, learning levels that lay down the conditions of possibility for learning processes in the dimensions of both objectivating thought and practical insight. I cannot go into that here. In any case, for the legitimation problems of the modern period, what is decisive is that the level of justification has become reflective. The procedures and presuppositions of justification are themselves now the legitimating grounds on which the validity of legitimations is based. The idea of an agreement that comes to pass among all parties, as free and equal, determines the procedural type of legitimacy of modern times. (By contrast, the classical type of legitimacy was determined by the idea of teachable knowledge of an ordered world.) Corresponding to this is an alteration of the position of the subject. Myth was taken for true in a naive attitude. The ordo-knowledge of God, the Cosmos, and the world of man was recognizable as the handed-down teachings of wise men or prophets. Those who make agreements under idealized conditions have taken the competence to interpret into their own hands.[17]

The procedural type of legitimacy was first worked out by Rousseau. The *contrat social* that seals the break with nature means a new principle of regulating behavior: the social. It shows by what path "justice can replace instinct in (human) behavior." That situation in which every individual totally gives himself and all his quasi-natural rights over to the community sums up the conditions under which only those regulations count as legitimate which express a common interest, that is, the general will: "For if each gives himself over completely, the situation is the same for all; and if the situation is the same for all, no one has an interest in making it difficult for others."[18] Of course, Rousseau did not understand his ideal contract only as the definition of a level of justification; he mixed the introduction of a new principle of legitimation with proposals for institutionalizing a just rule. The *volonté générale* was supposed not only to explicate grounds of validity but also to mark the place of sovereignty. This has confused the discussion of democracy right up to the present day.

I am thinking for one thing of the discussion of council democracy.[19] If one calls democracies precisely those political orders that satisfy the procedural type of legitimacy, then questions of democratization can be treated as what they are: as organizational questions. For it then depends on the concrete social and political conditions, on scopes of disposition, on information, and so forth, which types of organization and which mechanisms are in each case better suited to bring about procedurally legitimate decisions and institutions. Naturally one must think here in process categories. I can imagine the attempt to arrange a society democratically only as a self-controlled learning process. It is a question of finding arrangements which can ground the presumption that the basic institutions of the society and the basic political decisions would meet with the unforced agreement of all those involved, if they could participate, as free and equal, in discursive will-formation. Democratization cannot mean an a priori preference for a specific type of organization, for example, for so-called direct democracy.

The discussion between representatives of a normative theory of democracy on the one side and those of a "realistic" or empirical concept of democracy on the other has gone just as badly.[20] If democracies are distinguished from other systems of domination by a rational principle of legitimation and not by types of organization marked out a priori, then the opposing critics are missing their targets. Schumpeter and his followers reduce democracy to a method for selecting elites. I find this questionable, but not because, say, this competition of elites is incompatible with forms of basic democracy—one could imagine initial situations in which competitive-democratic procedures would be most likely to produce institutions and decisions having a presumption of rational legitimacy. I find Schumpeter's concept questionable because it defines democracy by procedures that have nothing to do with the procedures and presuppositions of free agreement and discursive will-formation. The procedures of democratic domination by elites are understood in a decisionistic manner, so that they cannot be linked with the idea of justification on the basis of generalizable interests. On the other side, normative theories of democracy are not to be faulted for holding fast to this procedural legitimacy. But they expose themselves to justified criticism as soon as they confuse a level of justification of domination with procedures for the organization of domination. If these are not kept separate, one can easily object—what Rousseau already knew —that there never was and never will be a true democracy.

The distinction between grounds for the validity of domination and institutions of domination evidently raises difficulties in view of the modern state. P. von Kielmannsegg, for instance, believes that agreement and consent may be made conditions for the legitimate exercise of domination; but they cannot be the ground of legitimacy because legitimacy arises only through "recourse to what is unconditionally valid." [21] Kielmannsegg thereby misses the modern point of the transposition of legitimate power to a reflective level of justification. Now only the procedures and presuppositions of agreement enjoy unconditional validity; an agreement counts as rational, that is, as an expression of a general interest, if it could only have come to pass under the ideal conditions that alone create legitimacy. A similar misunderstanding is present in the remarks made by Wilhelm Hennis [who spoke prior to Habermas at the meeting for which this paper was prepared]. According to him the legitimacy of the exercise of domination in the modern state rests on "penultimate grounds"; on this construction the term "ultimate grounds" would signify only limits of legitimate domination. Hennis is thinking, of course, of the privatization of the powers of faith with which the Wars of Religion were ended and of all that today sails under the flag of pluralism (a flag that conceals more than it reveals). But what did the religious neutralization of the state legitimize, if not (among other things) the discourses that were carried out from Hobbes through Hegel, that is, arguments which provided reasons or grounds [for holding] that such regulations were in the interest of everyone involved? Today it is neither ultimate nor penultimate grounds that provide legitimation. Whoever maintains this is operating at the level of the Middle Ages. Only the rules and communicative presuppositions that make it possible to distinguish an accord or agreement among free and equals from a contingent or forced consensus have legitimating force today. Whether such rules and communicative presuppositions can best be interpreted and explained with the help of natural law constructions and contract theories or in the concepts of a transcendental philosophy or a pragmatics of language or even in the framework of a theory of the development of moral consciousness is secondary in the present context.

The modern level of justification is also misconstrued by those who feel themselves to be above all that is old-European. They believe that a replacement for procedural legitimacy, in the sense of rational agreement, can be created through "procedure" in the sense of formal properties of the exercise of domination.[22] To be sure, the normative power of the factual is no chimera; but it is an indication that many norms have been established against the wills of those who follow them. Before norms of domination could be accepted *without reason* by the bulk of the population, the communication structures in which our motives for action have till now been formed would have to be thoroughly destroyed. Of course, we have no metaphysical guarantee that this will not happen.[23]

III

I would like now to take a brief look at the legitimation problems that emerge with the modern state. We characterize this state by features like the monopolization of legitimate power, centralized and rational (in Weber's sense) administration, territoriality, and so on. These features describes a structure of state organization that becomes visible only when we leave behind a narrowly political view fixated on the state and consider the emergence of capitalist society. This society requires a state organization different from that of the class societies of the great empires, which were constituted in an immediately political way (in ancient

Egypt, China, and India, as well as in European feudalism). Let me separate the internal and external aspects of this process.

Internally the modern state can be understood as the result of the differentiation of an economic system which regulates the production process through the market—that is, in a decentralized and unpolitical manner. The state organizes the conditions under which the citizens, as competing and strategically acting private persons, carry on the production process. The state itself does not produce, except perhaps as a subsidiary to entrepreneurs for whom certain functionally necessary investments are not yet or no longer profitable. In other words, the state develops and guarantees bourgeois civil law, the monetary mechanism, and certain infra-structures—overall the prerequisites for the continued existence of a depoliticized economic process set free from moral norms and orientations to use value. Since the state does not itself engage in capitalist enterprise, it has to siphon off the resources for its ordering achievements from private incomes. The modern state is a state based on taxation (Schumpeter). From these deter-minations there results a constellation of state and civil society which the Marxist theory of the state has been continually con-cerned to analyze.[24]

In comparison to the state of feudalism or the ancient empires, the modern state gains greater functional autonomy; the ability of the modern administration to assert itself vis-à-vis citizens and particular groups also grows in the framework of stronger func-tional specification. On the other hand, however, the comple-mentary relationship to the economy into which the state now enters makes clear for the first time the economic limitation on the state's scope of disposition. "Because (the state) is excluded from capitalist production as well as simultaneously dependent on it . . . it is forced to create the formal and (increasingly) also the material conditions and presuppositions for carrying on pro-duction and accumulation and for ensuring that their continuity does not founder on the material, temporal, and social instabilities inherent in the anarchic adaptation of the capital process to society."[25] The premodern state also faced the task of protecting society from disintegration without being able freely to dispose of the capacities for social integration; but the modern state directs its ordering achievements to delimiting a subsystem from its domain of sovereignty, a subsystem that replaces (at least in part) the social integration accomplished through values and norms with a system integration operating through exchange relations.[26]

As to the *external aspect* of the new state structure, the modern state did not emerge in the singular but as a system of states. It took shape in the Europe of the sixteenth century, where traditional power structures were dissipated and cultural homogeneity was rather great, where secular and spiritual authority had parted ways, trade centers had developed, and so on.[27] Wallerstein has shown that the modern system of states emerged in the midst of a "European world economy," that is, of a world market dominated by the European states.[28] The power differential between the centers and the periphery did not mean, however, that any single state had gained the power to control the worldwide relations of exchange. This means that the modern state took shape not only together with an internal economic environment but with an external one as well. This also explains the peculiar form of state sovereignty that is defined by relation to the sovereignty of other states. The private autonomy of individual, strategically acting, economic subjects is based on a reciprocal recognition that is secured legally and can be regulated universalistically. The political autonomy of individual, strategically acting, state powers is based on a reciprocal recognition that is sanctioned by the threat of military force and thus, despite agreement in international law, remains particular and quasi-natural. War and the mobilization of resources for building up standing armies and fleets are constitutive for the modern state system as it has existed for almost three hundred years since the Peace of Westphalia. The construction of a tax administration, of a central administrative apparatus in general, was at least as strongly shaped by this imperative as directly by the organizational needs of the capitalist economy.[29]

If one keeps these two aspects of the state structure before one's eyes, it becomes clear that the process of state building had to react upon the form of collective identity. The great empires were characterized by the fact that, as complex unities with a claim to universality, they could demarcate themselves externally—from a periphery territorially not precisely determined—only through incorporation, tributary subjugation, and association. The identity of such empires had to be anchored internally in the consciousness of only a small elite; it could coexist with other loosely integrated, prestate identities of archaic origin. The emergence of nations shows how this kind of collective identity was transformed under the pressure of the modern state structure. The nation is a (not yet adequately analyzed) structure of consciousness that satisfies at least two imperatives. First, it makes the formally egalitarian structures of bourgeois civil law (and later of political democracy) in internal relations subjectively compatible with the particularistic structures of self-assertion of sovereign states in external relations. Second, it makes possible a high degree of social mobilization of the population (for all share in the national consciousness). The French Revolution provides a model case for this; the nation emerged along with the bourgeois constitutional state and universal conscription.

I have recalled the structures of state and nation building because they can help decode the legitimation themes that accompanied the formation of the bourgeois state. If for the sake of simplicity we restrict ourselves to controversies concerning the theory of the state, we can (very roughly) distinguish five complexes.[30] These thematic strata run through several centuries. The first two reflect the constitution of the new level of justification, the other three the structures of the modern state and the nation.

a. *Secularization.* With the functional specification of the tasks of public administration and government, there developed a concept of the political that called for politically immanent justification. Detaching the legitimation of state power from religious traditions thus became a controversy of the first order. So far as I can see, Marsilius of Padua (drawing on Aristotle in his *Defensor Pacis* of 1324) was one of the first, if not the first, to criticize the theory of *translatio imperii* and thereby all theological justification.[31] This controversy extended into the nineteenth century, when conservative theoreticians such as De Bonald and De Maistre once again sought to ground religiously the traditional powers of church, monarchy, and a society of estates.

b. *Rational Law.* The great controversy between rational natural law and classical natural law, the effects of which also reached into the nineteenth century, focused on working out a procedural type of legitimation.[32] From Hobbes to Rousseau and Kant the leading ideas of rational agreement and self-determination were explicated to the extent that questions of justice and public welfare were stripped of all ontological connotations. This controversy dealt implicitly with the depreciation of a level of justification dependent on world views.

c. *Abstract Right and Capitalist Commodity Exchange.* Rational natural law had, of course, not only a formal side but a material side as well. From Hobbes and Locke through the Scottish moral philosophers (Hume, Smith, Millar), the French Enlightenment philosophers (Helvetius, d'Holbach), and classical political economy to Hegel, there emerged a theory of civil society that explained the bourgeois system of civil law, the basic liberties of the citizen, and the capitalist economic process as an order that guaranteed freedom and maximized welfare.[33] At the new level of justification only an order of state and society organized along universalistic lines could be defended. The controversy with the traditionalists concerned the historical price exacted by bourgeois ideals; it concerned, that is, the rights of the particular, the limits of rationality—from the perspective of the present, the "dialectic of enlightenment."

d. *Sovereignty.* The establishment of monarchic sovereignty within and without ignited a conflict that was carried out first along the fronts of the wars of religion. (See the political journalism of the Protestants after St. Bartholomew's Night in 1572.) From Bodin to Hobbes the sovereignty question was then resolved in favor of absolutism. In the course of the eighteenth century there was an attempt to rethink princely sovereignty into sovereignty of the people, so that the external sovereignty of the state could be unified with political democracy. The sovereignty of the people was, of course, a diffuse battle cry, which was unfolded in the constitutional debates of the nineteenth century. In it various thought motifs flow together: the sovereign power of the state appears as the expression of a new principle of legitimation, of the domination of the third estate, and of national identity as well.

e. *Nation.* This last complex has a special place insofar as national consciousness developed inconspicuously in very differentiated cultures, often on the basis of a common language, before it was dramatized in independence movements. Actually national identity became a controversial theme (in the nineteenth century) only where modernization processes were delayed, as in the succession states of the empire dissolved in 1804. A nationalism that served, as in the Bismarckian empire, to separate out internal enemies—*Reichsfeinde* such as socialists, Poles, and Catholics—no longer reflected the legitimacy thematic of the bourgeois state in its formative period; rather it now reflected the legitimacy conflicts into which this state fell when it became clear that modern bourgeois society did not dissolve class structures but first gave them pure expression *as* socioeconomic class structures.[34] This shock became permanent in the face of the threat to legitimacy represented since the nineteenth century, by the international labor movement.

Up to this point we have discussed legitimation themes that emerged with the development of the capitalist mode of production and the establishment of the modern state. They are an expression of legitimation problems on a scale that remains hidden so long as one limits oneself, as Hennis does, to the few parapets of the class struggle, the few historically significant legitimation crises, to the bourgeois revolutions. The extent of what has to be legitimated can be surmised only if one contemplates the vestiges of the centuries long repressions, the great wars, the small insurrections and defeats, that lined the path to the modern state. I am thinking, for example, of the resistance to what modernization research calls "penetration" (the establishment of administrative power)—hunger revolts when the food supply broke down, tax revolts when public exploitation became unbearable, revolts against the conscription of recruits, and so forth. These local insurrections against the offshoots of the modern state trickled away in the nineteenth century.[35] They were replaced by social confrontations of artisans, industrial workers, the rural proletariat. This dynamic produced new legitimation problems. The bourgeois state could not rely on the integrative power of national consciousness alone; it had to try to head off the conflicts inherent in the economic system and channel them into the political system as an institutionalized struggle over distribution. Where this succeeded, the modern state took on one of the forms of social welfare state-mass democracy.

IV

4. In regard to the legitimation problems in developed capitalist societies, I would like to make only a few remarks concerning a) a fundamental conflict which today gives rise to legitimation

problems, b) restrictive conditions on problem resolution, and
c) two stages of delegitimation.

a) The expression "social welfare state-mass democracy" mentions two properties of the political system which are effective for
legitimation. On the one hand, it tells us that the opposition to
the system which emerged in the labor movement has been defused by regulated competition between political parties. Among
other things, this has institutionalized oppositional roles, formalized and rendered permanent the process of legitimation, period-
ized variations in legitimation and canalized the withdrawal of
legitimation in the form of changes of regime, and finally it has
involved everyone in the legitimation process as voting citizens.

On the other hand, threats to legitimacy can be averted only
if the state can credibly present itself as a social welfare state
which intercepts the dysfunctional side-effects of the economic
process and renders them harmless for the individual—after the
fact, through a system of social security which is supposed to
mediate the basic risks connected with weak positions in the
market, and before the fact, through a system of securing the
conditions of life that is supposed to function primarily by way
of equal-opportunity access to formal schooling. In mass democracies, fulfilling this social welfare state program is, if not the
foundation, at least a necessary condition of legitimacy; it presupposes of course an economic system relatively free of disturbances. Thus the state programmatically assumes the responsibility to make good deficiencies in the functioning of the economic
process. There is today no disagreement concerning the structural
risks built into developed capitalist economies. These have to do
primarily with interruptions of the accumulation process conditioned by the business cycle, the external costs of a private production that cannot adequately deal with the problem situations
it itself creates, and a pattern of privilege whose core is a structurally conditioned unequal distribution of wealth and income.

The three great areas of responsibility against which the performance of the government is today measured are then: shaping
a business policy that ensures growth, influencing the structure
of production in a manner oriented to collective needs, and cor-
recting the pattern of social inequality. The problem does not lie

in the fact that such things are expected of the state and that the state has to take them up in a programmatic way. The conflict —in which, with Claus Offe, we can see a source of legitimation problems—lies rather in the fact that the state is supposed to perform all these tasks without violating the functional conditions of a capitalist economy, and this means without violating the complementarity relations that exclude the state from the economic system and, at the same time, also make it dependent on the dynamic of the economy.[36]

Viewed historically, the state was from the beginning supposed to protect a society determined normatively in its identity from disintegration, without ever having at its free disposal the capacities for social integration, without ever being able, as it were, to make itself master of social integration. The modern state at first fulfilled this function by guaranteeing the prerequisites for the continued existence of a private economic system free of the state. Disturbances and undesired side effects of the accumulation process did not have to result in the withdrawal of legitimation so long as the interests harmed could count as private interests and be segmented. To the extent, however, that the capitalist economic process penetrated ever broader areas of life and subjected them to its principle of societal adaptation, the systemic character of bourgeois society was consolidated. The interdependence of conditions in these once-private domains increased the susceptibility to disturbance and also gave these disturbances a politically relevant scale. Thus the dysfunctional side effects of the economic process could less and less be segmented from one another and neutralized in relation to the state. From this there grew *a general responsibility of the state for deficiencies* and a presumption of its competence to eliminate them. This places the state in a dilemma. On the one hand, the definitions of deficiencies and the criteria of success in dealing with them arise in the domain of political goal-settings that have to be legitimated; for the state has to deploy legitimate power if it takes on the catalog of tasks mentioned above. On the other hand, in this matter the state cannot deploy legitimate power in the usual way, to push through binding decisions, but only to manipulate the decisions of others, whose private autonomy may not be violated. Indirect control is the answer to the dilemma, and the limits to the effectiveness of indirect control signal the persistence of this dilemma.[37]

The legitimation problem of the state today is not how to conceal the functional relations between state activity and the capitalist economy in favor of ideological definitions of the public welfare. This is no longer possible—at least not in times of economic crisis—and exposure by Marxism is no longer necessary. The problem consists rather in representing the accomplishments of the capitalist economy as, comparatively speaking, the best possible satisfaction of generalizable interests—or at least insinuating that this is so. The state thereby programmatically obligates itself to keep dysfunctional side effects within acceptable limits. In this assignment of roles, the state provides legitimating support to a social order claiming legitimacy.

b. The state can prove itself as an aid to legitimation only if it successfully manages the tasks it has programmatically taken on; and to a considerable extent this can be checked. The legitimation theme that is today in the foreground can thus be located on the line between technocracy theories and participation models. I shall not go into this now.[38] But I would like to mention a series of restrictive conditions under which the state today must deal with those of its tasks effective for legitimation.[39]

1. The complementarity relationship between state and economy results in a goal conflict, of which there is a broadly effective awareness, especially in downward phases of the business cycle; the conflict is between a policy of stability that has to adjust its measures to the independent, cyclical dynamic of the economic process and, on the other hand, a policy of reform meant to compensate for the social costs of capitalist growth, which policy requires investments irrespective of the business situation and of profit considerations.[40]

2. The development of the world market, the internationalization of capital and labor,[41] has also placed external limits on the national state's latitude for action. The problems that arise for developing countries from international stratification can, it is true, be segmented to the point where they do not react back upon the legitimation process in developed countries. But the consequences of interlacing national economies with one another (e.g., the influence of multinational corpo-

rations) cannot be neutralized. The need for coordination at a supra-national level cannot easily be satisfied as long as governments have to legitimate themselves exclusively in terms of national decisions and thus have to react to national developments that are extremely non-synchronous.

3. Until the middle of this century, national identity in advanced European countries was so strongly developed that legitimation crises could be headed off by nationalistic means if in no other way. Today there are growing indications not only that exhaustion has set in where national consciousness has been overstimulated, but that a process of erosion is underway in all the older nations. The disproportion between worldwide mechanisms of system integration (world market, weapons systems, news media, personal communication) and the localized social integration of the state may be a contributing factor. Today it is no longer so easy to separate out internal and external enemies according to national characteristics. Characteristics relating to opposition to the system serve as a substitute (as, e.g., in the recent "Radicals Decree" in the Federal Republic of Germany). Conversely, however, membership in the system cannot, it seems, be built up to a positive identifying characteristic.

4. Nor are the sociostructural conditions particularly favorable for the "planning of ideology" (Luhmann). Horizontal and vertical expansion of the educational system does make it easier to exercise social control through the mass media. But the symbolic use of politics (in the sense of M. Edelmann) thereby also becomes more and more susceptible to exercises in self-contradiction. (On the evening news, one watches the leaders of the Social Democratic Party spell out investment control as a "forward-looking industrial policy." The next day one reads Wehner's denial in *Der Spiegel:* "We live at a time in which semantics is decisive." I am ignoring for the moment the selective circulation of *Der Spiegel*.)

c. If under these restrictive conditions the state does not succeed in keeping the dysfunctional side effects of the capitalist economic process within bounds acceptable to the voting public, if it is also unsuccessful in lowering the threshold of acceptability itself, then manifestations of delegitimation are unavoidable. This is marked at first by symptoms of a sharpened struggle over distribution, which proceeds according to the rules of a zero-sum game between the state's share, the share of wages, and the rate of profit. The rate of inflation, the financial crisis of the state, and the rate of unemployment—which can be substituted for one another only within limits—are measures of failure in the tasks of securing stability; the breakdown of reform politics is a sign of failure in the task of altering undesirable structures of production and privilege. At the moment, some of these symptoms can be found in the Federal Republic of Germany; yet the repercussions in the political sphere are almost minimal. I do not have data at my disposal with which we might satisfactorily explain this situation, and which would allow us to make a correct estimate of the weight of particular factors—for example, the role of a turn-around, emanating from the universities, which was consciously brought about through the mobilization of fear, much anthropological pessimism, adjuration of the virtues of subordination, and little argument.

Delegitimations on this level presuppose that the categories of rewards over which the distributional struggle is carried on are not themselves contested. One wants money, free time, and security. These "primary goods" are represented as neutral means for attaining an indefinite multiplicity of concrete ends selected according to values. These are certainly highly abstract means that can be employed for a number of purposes; nevertheless, these media lay down clearly circumscribed "opportunity structures." In them is reflected a form of life, the form of life of private commodity owners who bring their property—labor power, products, or means of payment—into exchange relations and thereby accommodate the capitalist form of mobilizing resources.[42] I shall not go through the characteristics of this familial, vocational, and civil privatism in detail. Nor shall I criticize the form of life that has its crystallizing point in possessive individualism (McPherson). I doubt, however, whether the form of life mirrored in system-conforming rewards can today—in the light of the alternatives opened by capitalist development itself—still be as convincingly legitimated as it could in Hobbes' time. Of course, such questions relevant to legitimation need not even be allowed if the powers that be are successful in further redefining practical questions into technical questions, if they are successful in preventing questions that radicalize the value-universalism of bourgeois society from even arising.

Otherwise the "pursuit of happiness" might one day mean something different—for example, not accumulating material objects of which one disposes privately, but bringing about social relations in which mutuality predominates and satisfaction does not mean the triumph of one over the repressed needs of the other. In this connection it is important whether the educational systems can again be coupled to the occupational system, and whether discursive desolidification of the (largely externally controlled or traditionally fixed) interpretations of our needs in homes, schools, churches, parliaments, planning administrations, bureaucracies, in culture production generally, can be avoided.

V

In closing I would like to return to the conceptual-analytic starting point of our reflections. What is the significance of the reconstructive concept I am using in analyzing legitimation problems?

The treatment of legitimation problems by social scientists, including Marxist theoreticians,[43] today moves in Max Weber's "sphere of influence." The legitimacy of an order of domination is measured against the *belief* in its legitimacy on the part of those subject to the domination. This is a question of the "belief that the structures, procedures, actions, decisions, policies, officials, or political leaders of a state possess the quality of rightness, of appropriateness, of the morally good, and ought to be recognized in virtue of this quality."[44] For systems theory (Parsons, Easton, Luhmann) this poses the question: With the help of which mechanisms can an adequate supply of legitimation be created, or through which functional equivalents can missing legitimation be replaced?[45] Learning theorists accommodate the question of the sociopsychological conditions under which a belief in legitimacy arises in a theory of the motivation for obedience.[46] Thus the empiricist replacement of legitimacy with what is held to be such allows for meaningful sociological investigations (the value of which will be decided by the success of the systems-theoretic and behaviorist approaches generally).

But we may well want to ask what price the empiricist must

pay for the redefinition of his object. If the object domain is conceived in such a way that not legitimate orders but only orders that are *held to be legitimate* can belong to it, then the connection between reasons and motives that exists in communicative action is screened out of the analysis. At least any independent evaluation of reasons is methodically excluded—the researcher himself refrains from any systematic judgment of the reasons on which the claim to legitimacy is based. Since the days of Max Weber this has been regarded as a virtue; however, even if one adopts this interpretation, the suspicion remains that legitimacy, the belief in legitimacy, and the willingness to comply with a legitimate order have something to do with motivation through "good reasons." But whether reasons are "good reasons" can be ascertained only in the performative attitude of a *participant* in argumentation, and not through the neutral *observation* of what this or that participant in a discourse holds to be good reasons. To be sure, the sociologist is concerned with the facticity of validity claims—for example, with the fact that the claim to legitimacy raised on behalf of a political order is recognized with specific frequencies in specific populations. But can he ignore the fact that normative validity claims meet with recognition because, among other reasons, they are held to be capable of discursive vindication, to be right, that is, to be well grounded? It is as with truth claims; the universality of this claim gives a sociologist the possibility of systematically checking the truth of an assertion independently of whether or not it is held to be true in a specific population. It can be decisive for an analysis to know whether a population acted on the basis of an accurate or a false opinion (e.g., to determine whether cognitive errors or other causes were principally responsible for observed failures). The case could be the same with the normative validity claim of political institutions; for example, one might well want to know whether a certain party renounces obedience because the legitimacy of the state *is* empty, or whether other causes are at work. To make that judgment we have to be able systematically to evaluate legitimacy claims in a rational, intersubjectively testable way. Can we do this?

Hennis is apparently of the opinion that we can. He holds a

"critical-normative demarcation of legitimacy from illegitimacy" to be absolutely necessary. But he does not specify the procedures or the criteria for this demarcation. He mentions legitimacy factors: personal esteem, efficiency in managing public tasks, approval of structures. But this personal authority is supposed to "spring from sources that can't be rationally grounded." What can count as efficient task management is measured against standards. These in turn are connected with those structures about whose legitimacy Hennis says only that they establish themselves in different national variants. He does not say what can count as a ground for the legitimacy of domination. To do so requires a concept of legitimacy with normative content. Hennis does not present us with such a concept, but he must have one, at least implicitly, in mind. The old-European style to his strategy of argumentation leads me to suspect connections with the classical doctrine of politics.

In this tradition (which goes back to Plato and Aristotle) stand important authors who still possess a substantial concept of morality, normative concepts of the good, of the public welfare, and so on.[47] Neo-Aristotelianism in particular experienced a renaissance in the writings of Hannah Arendt, Leo Strauss, Joachim Ritter, and others. The very title under which Ritter published his studies of Aristotle, "Metaphysics and Politics," suggests the difficulty of their position. Classical natural law is a theory dependent on world views. It was still quite clear to Christian Wolff at the the end of the eighteenth century that practical philosophy "presupposes in all its doctrines ontology, natural psychology, cosmology, theology, and thus the whole of metaphysics."[48] The ethics and politics of Aristotle are unthinkable without the connection to physics and metaphysics, in which the basic concepts of form, substance, act, potency, final cause, and so forth are developed. As Ritter puts it, in the *polis* that which is "by nature right" is realized because "with the *polis* the nature of man comes to its realization . . . whereas otherwise, where there is no *polis,* man can exist as man only in possibility but not in actuality."[49] Today it is no longer easy to render the approach of this metaphysical mode of thought plausible. It is no wonder that the neo-Aristotelian writings do not contain systematic doctrines, but are works of high interpretive art that suggest the truth of classical texts through interpretation, rather than by grounding it.

Thus certain reductive forms of Aristotelianism have a better chance. Withdrawing the theoretical claim of practical philosophy, they reduce it to a hermeneutics of everyday conceptions of the good, the virtuous, and the just in order then to certify that an unchangeable core of substantial morality is preserved in the prudent application of this knowledge. An example is Hennis' use of Aristotle's *Topics* for political science; another is Gadamer's interpretation of the *Nicomachean Ethics:*

Philosophical ethics is in the same situation in which everyone finds himself. What counts as right, what we consent or object to in judging ourselves or others, follows our general ideas of what is good and just; but it acquires genuine determinateness only in the concrete reality of the case, which is not a case of applying a general rule. . . . The general, the typical, which alone can be said in a philosophical investigation given over to the generality of the concept, is not essentially different from what guides the wholly untheoretical, average general consciousness of norms in every practical-moral reflection. Above all it is not different from it insofar as it includes the same task of application to given circumstances that belongs to all moral knowledge, that of the individual no less than that of the statesman acting in everyone's behalf.[50]

But if philosophical ethics and political theory can know nothing more than what is anyhow contained in the everyday norm consciousness of different populations, and if it cannot even know this in a different way, it cannot then rationally [*begründet*] distinguish legitimate from illegitimate domination. Illegitimate domination also meets with consent, else it would not be able to last. (One need only recall those days in which great masses of people came together in the squares and on the streets, without being pressured to do so, in order to acclaim an empire, a people, and a leader—was that an expression of anything other than an untheoretical, average norm consciousness?) If, on the other hand, philosophical ethics and political theory are supposed to disclose the moral core of the general consciousness and to *reconstruct* it as a normative concept of the moral, then they must specify criteria and provide reasons; they must, that is, produce theoretical knowledge.

An interesting language-analytic variant (inspired by the work of Wittgenstein) of the same embarrassment can be found in Hannah Pitkin's: *Wittgenstein and Justice*. She offers an interpretation of the dialogue on justice between Socrates and the Sophist Thrasymachus, which Plato reports in the first book of the *Republic*.[51] If we view Thrasymachus from the perspective of contemporary discussion, he represents an empiricist standpoint; for him justice is just another name for the particular interests of the stronger. Socrates develops a normative concept of justice; whoever calls something unjust must apply standards and be able to ground them as well. Both start from the fact that a great discrepancy had arisen between the normative content of the concept "justice" as it was then understood by the Greeks and the contemporary institutions, actions, and practices, that were supposed to be legitimate and to embody justice. But Socrates turns the concept critically against the institutions, while Thrasymachus deflates the concept for the purpose of describing behavior practiced in the name of justice.

Pitkin shows the difference between the grammars of the two language games, in which the same term is used, in one case with quotation marks, in the other without. We assume different "grammatically" regulated attitudes when we say, on the one hand, "I like the picture," and on the other, "The picture is beautiful" (for in the second case we can continue: "and yet I don't like it"). The situation is analogous when we say: "X fought for a just cause," and on the other hand: "X claimed to be fighting for a just cause" (for in the latter case we can continue: "but he was actually pursuing his own interests"). The attitude we assume in employing normative concepts like justice, beauty, and truth (with which universal validity claims are connected) is evidently deeply rooted in human forms of life; a change of attitude to the neutral position of the observer has to alter the meaning of these terms. But what follows from this for a reconstruction of the validity claims and normative content of the concepts in question? According to Pitkin: "Our concepts are conventional, but the conventions on which these concepts rest are not arbitrary; they are shaped by our human condition and conduct, by our forms of life." [52] That may well be. But who guarantees that the grammar of these forms of life not only regulates *customs* but gives expression to *reason*. It is only a small step from this conservative appropriation of the great traditions in terms of language games to the traditionalism of a Michael Oakeshott.[53] This is also the position that Hennis adopts when, while presupposing virtue and justice as the validity basis of legitimate domination, he nevertheless has recourse only to customs.

I have discussed two concepts of legitimation, the empiricist and the normativist. One can be employed in the social sciences but is unsatisfactory because it abstracts from the systematic weight of grounds for validity; the other would be satisfactory in this regard but is untenable because of the metaphysical context in which it is embedded. I would like, therefore, to propose a third concept of legitimation, which I shall call the "reconstructive."

I shall begin by assuming that the proposition: "Recommendation X is legitimate" has the same meaning as the proposition: "Recommendation X is in the general (or public) interest," where X can be an action as well as a norm of action or even a system of such norms (in the case we are considering, a system of domination). "X is in the general interest" is to mean that the normative validity claim connected with X counts as justified.[54] The justifiability of competing validity claims is decided by a system of possible justifications; a single justification is called a legitimation. The reconstruction of given legitimations can consist, first, in discovering the justificatory system, S, that allows for evaluating the given legitimations as valid or invalid in S. "Valid in S" is to mean only that anyone who accepts S—a myth or a cosmology or a political theory—must also accept the grounds given in valid legitimations. This necessity expresses a consistency connection resulting from the internal relations of the justificatory system.

In taking a justification up to this threshold, we have interpreted a belief in legitimacy and tested its consistency. Along this hermeneutic path alone, however, we do not arrive at a judgment of the legitimacy that is believed in. Nor does comparison of the belief in legitimacy with the institutional system justified take us much further. Assuming that idea and reality do not split apart, what is needed is rather an evaluation of the reconstructed justificatory system itself. This brings us back to the fundamental question of practical philosophy. In modern times it has been taken up reflectively as a question of the procedures and presuppositions under which justifications can have the power to produce consensus. I have mentioned the theory of justice of John Rawls, who examines how the original situation would have to be constituted so that a rational consensus about the basic decisions and basic institutions of a society could come to pass. Paul Lorenzen examines the methodic norms of the speech practice that makes rational consensus possible in such practical questions. Finally Karl-Otto Apel radicalizes this question with regard to the universal and necessary—that is, transcendental—presuppositions of practical discourse; the normative content of the universal presuppositions of communication is supposed thereby to form the core of a universal ethics of speech.[55] This is the point of convergence toward which the attempts to renew practical philosophy today seem to strive.

Even if we assent to this thesis, an objection comes readily to mind. Every general theory of justification remains peculiarly abstract in relation to the historical forms of legitimate domination. If one brings standards of discursive justification to bear on traditional societies, one behaves in an historically "unjust" manner. Is there an alternative to this historical injustice of general theories, on the one hand, and the standardlessness of mere historical understanding, on the other? The only promising program I can see is a theory that structurally clarifies the historically observable sequence of different levels of justification and reconstructs it as a developmental-logical nexus.[56] Cognitive developmental psychology, which is well corroborated and which has reconstructed ontogenetic stages of moral consciousness in this way, can be understood at least as a heuristic guide and an encouragement.

Notes

Notes to Translator's Introduction

1. Cf. "Literaturbericht zur philosophischen Diskussion um Marx und den Marxismus," in *Philosophische Rundschau* 5 (1957):165-235; reprinted in *Theorie und Praxis* (Neuwied, 1963), pp. 261-335.
2. Cf. "Between Philosophy and Science: Marxism as Critique," in *Theory and Practice* (Boston, 1973), pp. 195-252.
3. Cf. "The Classical Doctrine of Politics in Relation to Social Philosophy," in *Theory and Practice*, pp. 41-81.
4. Ibid., p. 79.
5. "Literaturbericht," p. 310.
6. "Literaturbericht zur Logik der Sozialwissenschaften," *Philosophische Rundschau* Beiheft 5 (1967); reprinted in *Zur Logik der Sozialwissenschaften* (Frankfurt, 1970), pp. 71-310; *Erkenntnis und Interesse* (Frankfurt, 1968), English translation, *Knowledge and Human Interests* (Boston, 1971).
7. *Zur Logik der Sozialwissenschaften*, pp. 251-290; English translation, "A Review of Gadamer's *Truth and Method*," in F. Dallmayr and T. McCarthy, eds., *Understanding and Social Inquiry* (Notre Dame, 1977), pp. 335-363.
8. *Knowledge and Human Interests*, chaps. 10-12.
9. "Der Universalitätsanspruch der Hermeneutik," in K. O. Apel et al., *Hermeneutik und Ideologiekritik* (Frankfurt, 1971), p. 138.
10. *Knowledge and Human Interests*, p. 269.
11. *Zur Logik der Sozialwissenschaften*, pp. 164-184, 305-308; cf. also "Theorie der Gesellschaft oder Sozialtechnologie? Eine Auseinandersetzung mit Niklas Luhmann," in *Theorie der Gesellschaft oder Sozialtechnologie—Was leistet die Systemforschung?* (Frankfurt, 1971), pp. 142-290.
12. *Zur Logik der Sozialwissenschaften*, pp. 181-182.

13. Ibid., pp. 176–177.

14. Cf. the *Vorwort* to the 1970 edition of *Zur Logik der Sozialwissenschaften* in which Habermas warns against confusing "processes of self-understanding with their results," and states that he would then (1970) develop the "discussion fragments" presented there in another direction. In the "Postscript to *Knowledge and Human Interests*," *Philosophy of the Social Sciences* 3 (1973):157–189, he reminds us that this was intended only as a prolegomenon (p. 159); although he still wants "to uphold the systematic conception of the book, . . . this idea assumes a somewhat different complexion" when the necessary refinements have been worked out (p. 158).

15. Published as the Appendix to *Knowledge and Human Interests*, pp. 301–317; here p. 314.

16. With one important exception: the idea of discourse and the conception of an "ideal speech situation" that is connected with it; cf. "Vorbereitende Bemerkungen zu einer Theorie der kommunikativen Kompetenz," in *Theorie der Gesellschaft oder Sozialtechnologie*, pp. 101–141; "Wahrheitstheorien," in *Wirklichkeit und Reflexion: Festschrift für Walter Schulz* (Pfullingen, 1973), pp. 211–265; cf. also T. McCarthy, *The Critical Theory of Jürgen Habermas* (Cambridge, Mass., 1978), chaps. 4.2 and 4.3.

17. Cf. "Stichworte zur Theorie der Sozialisation," in *Kultur und Kritik* (Frankfurt, 1973), pp. 118–194; "Notizen zum Begriff der Rollenkompetenz," in ibid., pp. 195–231; "Zur Einführung," in R. Döbert, J. Habermas, and G. Nunner-Winkler, eds., *Die Entwicklung des Ichs* (Köln, 1977); cf. also McCarthy, *The Critical Theory of Jürgen Habermas*, chap. 4.4.

18. "Geschichte und Evolution," in *Zur Rekonstruktion des Historischen Materialismus* (Frankfurt, 1976), pp. 200–259; here p. 250.

19. *Legitimation Crisis* (Boston, 1975).

Notes to "What Is Universal Pragmatics?"

1. Hitherto the term "pragmatics" has been employed to refer to the analysis of particular contexts of language use and not to the reconstruction of universal features of using language (or of employing sentences in utterances). To mark this contrast, I introduced a distinction between "empirical" and "universal" pragmatics. I am no longer happy with this terminology; the term "formal pragmatics"—as an extension of "formal semantics"—would serve better. *"Formalpragmatik"* is the term preferred by F. Schütze, *Sprache Soziologisch Gesehen*, 2 vols. (Munich, 1975); cf. the summary 911–1024.

2. I shall focus on an idealized case of communicative action, viz. "consensual interaction," in which participants share a tradition and their ori-

entations are normatively integrated to such an extent that they start from the same definition of the situation and do not disagree about the claims to validity that they reciprocally raise. The following schema locates the extreme case of consensual interaction in a system of different types of social action. Underlying this typology is the question of which categories of validity claims participants are supposed to raise and to react to.

These action types can be distinguished by virtue of their relations to the validity basis of speech:

a) *Communicative vs. Strategic Action.* In communicative action a basis of mutually recognized validity claims is presupposed; this is not the case in strategic action. In the communicative attitude it is possible to reach a direct understanding oriented to validity claims; in the strategic attitude, by contrast, only an indirect understanding via determinative indicators is possible.

b) *Action Oriented to Reaching Understanding vs. Consensual Action.* In consensual action agreement about implicitly raised validity claims can be *presupposed* as a background consensus by reason of common definitions of the situations; such agreement is supposed to be *arrived at* in action oriented to reaching understanding. In the latter case strategic elements may be employed under the proviso that they are meant to lead to a direct understanding.

c) *Action vs. Discourse.* In communicative action it is naively supposed that implicitly raised validity claims can be vindicated (or made immediately plausible by way of question and answer). In discourse, by contrast, the validity claims raised for statements and norms are hypothetically bracketed and thematically examined. As in communicative action, the participants in discourse retain a cooperative attitude.

d) *Manipulative Action vs. Systematically Distorted Communication.*

Whereas in systematically distorted communication at least one of the participants deceives *himself* about the fact that the basis of consensual action is only apparently being maintained, the manipulator deceives at least one of the *other* participants about his own strategic attitude, in which he *deliberately* behaves in a pseudoconsensual manner.

3. K.-O. Apel, "Sprechakttheorie und transzendentale Sprachpragmatik —zur Frage ethischer Normen," in K.-O. Apel, ed., *Sprachpragmatik und Philosophie* (Frankfurt, 1976), pp. 10–173.

4. In the framework of Southwest German Neo-Kantianism, Emil Lask earlier reconstructed the concept of "transsubjective validity"— in connection with the meaning of linguistic expressions, the truth of statements, and the beauty of works of art—as worthiness to be recognized. Lask's philosophy of validity combines motifs from Lotze, Bolzano, Husserl, and, naturally, Rickert. "Genuine value is worthiness to be recognized, recognition-value, that which deserves submission, that to which it is due, thus that which demands or requires it. To be valid is value, demand, norm. . . . All such terms as 'worthiness,' 'deserve,' 'be due,' 'demand' are correlative concepts; they point to a subjective behavior corresponding to validity—worthy to be treated or regarded in a certain way, it demands a certain behavior." E. Lask, "Zum System der Logik," *Ges. Schriften,* vol. 3 (Tübingen, 1924), p. 92.

5. Y. Bar-Hillel fails to appreciate this in his critique: "On Habermas' Hermeneutic Philosophy of Language," *Synthese* 26 (1973):1–12. His critique is based on a paper I characterized as provisional: "Vorbereitende Bemerkungen zu einer Theorie der kommunikativen Kompetenz," in J. Habermas and N. Luhmann, *Theorie der Gesellschaft oder Sozialtechnologie* (Frankfurt, 1971), pp. 101–141. Bar-Hillel has, I feel, misunderstood me on so many points that it would not be fruitful to reply in detail. I only hope that in the present sketch I can make my (still strongly programmatic) approach clear even to readers who are aggressively inclined and hermeneutically not especially open.

6. E.g., K.-O. Apel, *Transformation der Philosophie,* vol. 2 (Frankfurt, 1971), pp. 406 ff., and "Programmatische Bemerkungen zur Idee einer transzendentalen Sprachpragmatik," in *Annales Universitatis Tukuensis Sarja,* Series B, Osa Tom 126 (Tuku, 1973), pp. 11–35.

7. Charles Morris, "Foundations of the Theory of Signs," in *Encyclopedia of Unified Science* vol. 1, no. 2 (Chicago, 1938), and *Signs, Language, Behavior* (New York, 1955).

8. Cf. my remarks on Morris in *Zur Logik der Sozialwissenschaften* (Frankfurt, 1970), pp. 150 ff.

9. Bar-Hillel, "Indexical Expressions," in *Aspects of Language* (Jerusalem, 1970), pp. 69–88, and "Semantics and Communication," in H. Heidrich, *Semantics and Communication* (Amsterdam, 1974), pp. 1–36. Taking Bar-Hillel as his point of departure, A. Kasher has proposed a formal representation embedding linguistic expressions in extralinguistic contexts.

"A Step Forward to a Theory of Linguistic Performance," in Y. Bar-Hillel, ed., *Pragmatics of Natural Languages* (Dordrecht, 1971), pp. 84–93; cf. also R. C. Stalnaker, "Pragmatics," in Davidson and Harman, *Semantics of Natural Language* (Dordrecht, 1972), pp. 380–387.

10. R. M. Hare, *The Language of Morals* (Oxford, 1952); G. H. von Wright, *Norm and Action* (London, 1963); and N. Rescher, *Topics in Philosophical Logic* (Dordrecht, 1968).

11. L. Apostel, "A Proposal on the Analysis of Questions," in *Logique et Analyse*, 12(1969):376–381; and W. Kuhlmann, *Reflexion zwischen Theorie und Kritik* (Frankfurt, 1975).

12. S. Toulmin, *The Uses of Argument* (Cambridge, 1964); W. C. Salmon, *The Foundation of Scientific Inference* (Pittsburg, 1967); and cf. the summary chapter on "non-demonstrative inference" in R. P. Botha, *The Justification of Linguistic Hypotheses* (The Hague, 1973), pp. 25–72.

13. F. Kiefer, "On Presuppositions," in F. Kiefer and N. Ruwet, eds., *Generative Grammar in Europe* (Dordrecht, 1973), pp. 218–242; K. H. Ebert, "Präsuppositionen im Sprechakt," in Cate and Jordens, eds., *Linguistische Perspektiven* (Tübingen, 1973), pp. 45–60; and F. Petöfi, *Präsuppositionen in Linguistik und Philosophie* (Frankfurt, 1974).

14. H. P. Grice, "The Logic of Conversation," unpubl. MS (1968); and D. Gordon and G. Lakoff, "Conversational Postulates," unpubl. MS (1973).

15. J. R. Ross, "On Declarative Sentences," in Jacob Rosenbaum, ed., *Readings in English Transformational Grammar* (Waltham, Mass., 1970), pp. 222–277; J. D. MacCawley, "The Role of Semantics in a Grammar," in Bach and Harms, *Universals of Language* (New York, 1968), pp. 125–170; and D. Wunderlich, "Sprechakte," in Mass and Wunderlich, *Pragmatik und sprachliches Handeln* (Frankfurt, 1972), pp. 69–188, and "Zur Konventionalität von Sprechhandlungen," in D. Wunderlich, ed., *Linguistische Pragmatik* (Frankfurt, 1972), pp. 11–58.

16. C. J. Fillmore, "Pragmatics and the Description of Discourse," unpubl. MS (1973); G. Posner, *Textgrammatik* (Frankfurt, 1973).

17. J. Lyons, *Introduction to Theoretical Linguistics* (New York, 1968); J. J. Katz, *Semantic Theory* (New York, 1972).

18. P. F. Strawson, *Logico-Linguistic Papers* (London, 1971).

19. A. C. Danto, *Analytical Philosophy of Action* (Cambridge, 1973); S. Hampshire, *Thought and Action* (London, 1960); D. S. Schwayder, *The Stratification of Behavior* (London, 1965); and Care and Landesman, eds., *Readings in the Theory of Action* (London, 1968).

20. P. Winch, *The Idea of a Social Science and Its Relation to Philosophy* (London, 1958); C. Taylor, "Explaining Action," *Inquiry* 13 (1973): 54–89; and H. von Wright, *Explanation and Understanding* (London, 1971), and "On the Logic and Epistemology of the Causal Relation," in P. Suppes, ed., *Logic, Methodology and Philosophy of Science*, vol. 4 (1973), pp. 239–312.

21. W. P. Alston, *Philosophy of Language* (Englewood Cliffs, N.J., 1964).

22. J. Bennett, "The Meaning-Nominalist Strategy," *Foundations of Language* 10 (1973):141–168; and S. R. Schiffer, *Meaning* (Oxford, 1972).

23. Cf. the bibliography by E. von Savigny in J. L. Austin, *Zur Theorie der Sprechakte* (Stuttgart, 1972), pp. 203 ff.

24. G. Grewendorf, "Sprache ohne Kontext," in D. Wunderlich, ed., *Linguistische Pragmatik*, pp. 144–182.

25. H. P. Grice, "Meaning," *Philosophical Review* 66(1957):377–388, and "Utterer's Meaning, Sentence Meaning, and Word Meaning," *Foundations of Language* 4 (1968):225–242; and D. K. Lewis, *Convention* (Cambridge, 1969).

26. J. Habermas, *Zur Logik der Sozialwissenschaften*, pp. 184 ff.

27. H. G. Gadamer emphasizes this in *Truth and Method* (New York, 1975).

28. G. Ryle, *The Concept of Mind* (London, 1949); cf. the interpretation of E. von Savigny in *Die Philosophie der normalen Sprache* (Frankfurt, 1974), pp. 91 ff.

29. R. Carnap, W. Stegmüller, *Induktive Logik und Wahrscheinlichkeit* (Wien, 1959), p. 15.

30. D. Wunderlich, *Grundlagen der Linguistik* (Hamburg, 1974), p. 209.

31. For an analysis of what explication in the sense of rational reconstruction means, cf. H. Schnädelbach, *Reflexion und Diskurs* (Frankfurt, 1977), the chapter on "Explikativer Diskurs," pp. 277–336.

32. N. Chomsky, *Aspects of the Theory of Syntax* (Cambridge, Mass., 1965).

33. Wunderlich, *Grundlagen*, pp. 210–218.

34. R. P. Botha, *Justification*, pp. 75 ff., speaks in this connection of external versus internal linguistic evidence.

35. Wunderlich, *Grundlagen*, p. 216. If I understand correctly, Schnelle gives an empirical interpretation to the model-theoretic version of linguistics in *Sprachphilosophie und Linguistik* (Hamburg, 1973), pp. 82–114.

36. Botha, *Justification*, p. 224, thinks that a speaker can not only report correct linguistic intuitions falsely but can also have false intuitions; but the construct of pretheoretical *knowledge* does not allow of this possibility. I think it makes sense to start from the idea that linguistic intuitions can be "false" only if they come from incompetent speakers. Another problem is the interplay of grammatical and nongrammatical (e.g., perceptual) epistemic systems in the formation of diffuse judgments about the acceptability of sentences, that is, the question of isolating expressions of grammatical rule-consciousness, of isolating genuinely linguistic intuitions.

37. In this connection, U. Oevermann points out interesting parallels with Piaget's concept of reflecting abstraction. Perhaps the procedure of rational reconstruction is only a stylized and, as it were, controlled form of the reflecting abstraction the child carries out when, for example, it "reads off" of its instrumental actions the schema that underlies them.
38. W. J. M. Levelt, *Formal Grammars in Linguistics and Psycholinguistics,* vols. 1–3 (Amsterdam, 1974).
39. Levelt, *Formal Grammars,* vol. 3, pp. 5–7.
40. Ibid., pp. 14 ff.
41. In response to the doubts that Botha raises against the "clear case principle" (*Justification,* p. 224), I would like to reproduce an argument that J. J. Katz and B. G. Bever have brought against similar doubts in a paper critical of empiricism, "The Fall and Rise of Empiricism," unpubl. MS (Feb. 1974), pp. 38–39:

Such a theory ... seeks to explicate intuitions about the interconnectedness of phonological properties in terms of a theory of the phonological component, to explicate intuitions about the interconnectedness of syntactic properties in terms of a theory of the syntactic component, and to explicate intuitions about the interconnectedness of semantic properties in terms of a theory of the semantic component. The theory of grammar seeks finally to explicate intuitions of relatedness among properties of different kinds in terms of the systematic connections expressed in the model of a grammar that weld its components in a single integrated theory of the sound-meaning correlation in a language.

These remarks are, of course, by way of describing the theoretical ideal. But as the theory of grammar makes progress toward this ideal, it not only sets limits on the construction of grammars and provides a richer interpretation for grammatical structures but it also defines a wider and wider class of grammatical properties and relations. In so doing, it marks out the realm of the grammatical more clearly, distinctly, and securely than could have been done on the basis of the original intuitions. As Fodor has insightfully observed, such a theory literally defines its own subject matter in the course of its progress: "There is then an important sense in which a science has to discover what it is about; it does so by discovering that the laws and concepts it produced in order to explain one set of phenomena can be fruitfully applied to phenomena of other sorts as well. It is thus only in retrospect that we can say of all the phenomena embraced by a single theoretical framework that *they* are what we meant, for example, by the presystematic term 'physical event,' 'chemical interaction,' or 'behavior.' To the extent that such terms, or their employments, are neologistic, the neologism is occasioned by the insights that successful theories provide into the deep similarities that underlie superficially heterogeneous events." [J. A. Fodor, *Psychological Explanation* (New York, 1968), pp. 10–11].

42. H. Leuninger, M. H. Miller, and F. Müller, *Psycholinguistik* (Frankfurt, 1973), and *Linguistik und Psychologie* (Frankfurt, 1974); H.

Leuninger, "Linguistik und Psychologie," in R. Bartsch and Vennemenn, eds., *Linguistik und Nachbarwissenschaften* (Kronberg, 1973), pp. 225–241.

43. E. H. Lenneberg, *Biologische Grundlagen der Sprache* (Frankfurt, 1972), and "Ein Wort unter uns," in Leuninger et al., *Linguistik und Psychologie,* pp. 53–72.

44. L. Kohlberg, "Stage and Sequence," in D. Goslin, ed., *Handbook of Socialization Theory and Research* (Chicago, 1969), and "From is to Ought," in T. Mischel, ed., *Cognitive Development and Epistemology* (New York, 1971), pp. 151–236.

45. On this point, cf. U. Oevermann, "Kompetenz und Peformanz," unpubl. MS, Max-Planck-Institut für Bildungsforschung (1974).

46. Kant, *Critique of Pure Reason,* N. Kemp Smith, trans. (New York, 1961), p. 138.

47. B. Stroud, "Transcendental Arguments," *Journal of Philosophy* 9 (1968):241–254; M. S. Gram, "Transcendental Arguments," *Nous* 5 (1971):15–26; J. Hintikka, "Transcendental Arguments," *Nous* 6 (1972): 174–281; and M. S. Gram, "Categories and Transcendental Arguments," *Man and World* 6 (1973):252–269.

48. R. Bittner, "Transzendental," in *Handbuch philosophischer Grundbegriffe,* vol. 5 (Munich, 1974), pp. 1524–1539.

49. For example, the Kant reception of the Erlangen school assumes a transcendental status for the basic concepts of protophysics only in a limited sense; cf. the discussion volume edited by G. Böhme, *Protophysik* (Frankfurt, 1975).

50. Piaget's Kantianism is typical of this approach.

51. Cf. K.-O. Apel's introductions to volumes 1 and 2 of C. S. Peirce's *Schriften* (Frankfurt, 1967 and 1970).

52. Cf. my "Postscript to *Knowledge and Human Interests,*" *Philosophy of the Social Sciences* 3(1973):157–189; cf. also R. Bubner, "Transzendentale Hermeneutik," in Simon-Schäfer and Zimmerli, eds., *Wissenschaftstheorie der Geisteswissenschaften* (Hamburg, 1975), pp. 57–70.

53. F. Kambartel, *Erfahrung und Struktur* (Frankfurt, 1968).

54. J. Habermas, "Wahrheitstheorien," in *Festschrift für Walter Schulz* (Pfullingen, 1973), pp. 211–265.

55. W. Sellars, "Presupposing," *Philosophical Review* 63 (1954):197–215; P. F. Strawson, "A Reply to Mr. Sellars," *Philosophical Review* 63 (1954):216–231.

56. U. Oevermann, "Theorie der individuellen Bildungsprozesse," unpubl. MS, Max-Planck-Institut für Bildungsforschung (1974).

57. On this point, cf. the controversy between Quine and Chomsky: N. Chomsky, "Quine's Empirical Assumptions," and W. V. O. Quine, "Replies," both in Davidson and Hintikka, eds., *Words and Objections* (Dordrecht, 1969), pp. 53–68 and 292–352; W. V. O. Quine, "Methodological Reflections on Current Linguistic Theory," in Davidson and

Harman, eds., *Semantics of Natural Language* (Dordrecht, 1972). H. Schnelle, *Sprachphilosophie und Linguistik* (Hamburg, 1973) is also typical of methodological behaviorism in linguistics.
58. J. L. Austin, *How to do Things with Words* (Oxford, 1962); cf. the bibliography on the theory of speech acts compiled by E. von Savigny for the German edition of this work (Stuttgart, 1972), pp. 204–209. J. L. Austin, "Performative Utterances," in *Philosophical Papers* (Oxford, 1970), pp. 233–252, and "Performative-Constative," in C. E. Caton, ed., *Philosophy and Ordinary Language* (Urbana, Ill., 1963), pp. 22–33. Additional Austin bibliography can be found in von Savigny, *Die Philosophie der normalen Sprache,* pp. 162–166.
See also J. R. Searle: "What is a Speech Act?," in M. Black, ed., *Philosophy in America* (Ithaca, 1965), pp. 221–239, reprinted in Rosenberg and Travis, eds., *Readings in the Philosophy of Language* (Englewood Cliffs, N.J., 1971), pp. 614–628; "Austin on Locutionary and Illocutionary Acts," *Philosophical Review* 77 (1968):405–424, reprinted in ibid., pp. 262–275; *Speech Acts* (London, 1969); and "Linguistik und Sprachphilosophie," in Bartsch and Vennemann, *Linguistik und Nachbarwissenschaften,* pp. 111–126.
Other sources include: W. P. Alston, *Philosophy of Language,* and "Linguistic Acts," *American Philosophical Quarterly* 1(1964):138–146; L. J. Cohen, "Do Illocutionary Forces Exist?," *Philosophical Quarterly* 14 (1964):118–137, reprinted in Rosenberg and Travis, *Readings,* pp. 580–598, and "Speech Acs," *Current Trends in Linguistics* 12 (1970); R. M. Hare, "Meaning and Speech Acts," *Philosophical Review* 79 (1970):3–24, and "Austin's Distinction between Locutionary and Illocutionary Acts," in Hare, *Practical Inferences* (London, 1972); D. Holdcroft, "Performatives and Statements," *Mind* 83 (1974):1–18; P. F. Strawson, "Intention and Convention in Speech Acts," *Philosophical Review* 73 (1964):439–460, reprinted in Rosenberg and Travis, *Readings,* pp. 599–613; S. Thau, "The Distinction between Ebetic and Illocutionary Acts," *Analysis* 32 (1972/73):177–183; C. Travis, "A Generative Theory of Speech Acts," in Rosenberg and Travis, *Readings,* pp. 629–644; G. J. Warnock, "Hare on Meaning and Speech Acts," *Philosophical Review* 80 (1971):80–84; Wunderlich, *Grundlagen der Linguistik,* pp. 309–352.
59. Chomsky, *Aspects of the Theory of Syntax,* pp. 3 ff.
60. These qualifications are stated below in the discussion of Searle's principle of expressibility.
61. P. F. Strawson, *Individuals* (London, 1959); M. Dummet, *Frege, Philosophy of Language* (London, 1973); E. Tugendhat, *Vorlesungen zur Einführung in die sprachanalytische Philosophie* (Frankfurt, 1976).
62. On the analysis of intentionality and the expression of intentions, cf. W. Sellars, "Empiricism and the Philosophy of Mind," in *Metaphysics* (London, 1968); W. Sellars and R. Chisholm, "Intentionality and the Mental," in *Minnesota Studies,* vol. 1 (1963), pp. 507–539; Sellars, *Sci-*

ence and Metaphysics (London, 1968); E. Tugendhat, "Phänomenologie und Sprachanalyse," in *Festschrift für Gadamer,* vol. 2 (Tübingen, 1970), pp. 3–24; J. Hintikka, *Knowledge and Belief* (Ithaca, 1962); C. Taylor, "Explaining Action," *Inquiry* 13 (1970):54–89. On the analysis of expressive speech acts, cf. P. M. S. Hacker, *Insight and Illusion* (Oxford, 1972), chs. 7, 8, 9.

63. Cf. B. B. Steinberg and Jakobovits, eds., *Semantics* (Cambridge, 1971), pp. 157–484; H. E. Boekle, *Semantik* (Munich, 1972).

64. The work of P. W. Alston is a good example.

65. F. von Kutschera, *Sprachphilosophie* (Munich, 1971), pp. 117–161; H. Schnelle, *Sprachphilosophie und Linguistik,* pp. 190–240; Wunderlich, *Grundlagen,* pp. 238–273.

66. P. Watzlawick, J. H. Beavin, D. D. Jackson, *Pragmatics of Human Communication* (New York, 1967).

67. A communication theory that is supposed to reconstruct conditions of action oriented to reaching understanding requires as its basic unit of analysis, not necessarily pairs of complementary speech actions—that is, reciprocally performed and accepted speech actions—but at least a speaker's utterance that can not only be understood but accepted by at least one additional speaking and acting subject.

68. D. Wunderlich, "Zur Konventionalität von Sprechhandlungen," in D. Wunderlich, ed., *Linguistische Pragmatik,* p. 16; cf. also the linguistic characterization of the standard form given there and Wunderlich's analysis of advising in *Grundlagen,* pp. 349 ff.

69. Exceptions are representative speech acts, which, when rendered explicit, can also take on a negative form; e.g., "I do not want (hereby) to conceal from you that . . ."

70. Deviating from a common usage, I do not think it advisable to distinguish propositions from assertions in such a way that a proposition is indeed embedded in a specific speech situation through being asserted but does not receive its assertoric force therefrom. I am of the opinion, rather, that the assertoric force of a proposition cannot be reconstructed otherwise than with reference to the validity claim that anyone in the role of a competent speaker could raise for it in asserting it. Whether this claim can, if necessary, be discursively vindicated, that is, whether the proposition is "valid" (true), depends on whether it satisfies certain truth conditions. We can, to be sure, view propositions monologically, that is, as symbolic structures with an abstract truth value without reference to a speaker; but then we are abstracting precisely from the speech situation in which a propositional content, owing to the fact that it is asserted as a proposition, receives a relation to reality, that is, fulfills the condition of being true or false. This abstraction naturally suggests itself (and often remains hidden even from the logician) because the truth claim raised by the speaker is *universalistic*—that is, precisely of such a nature that, although it is raised in a particular situation, it could be defended against *anyone's* doubt.

71. S. Kanngiesser, "Aspekte zur Semantik und Pragmatik," in *Linguistische Berichte* 24 (1973):1–28, here p. 5.

72. Wunderlich, *Grundlagen,* pp. 337 ff.

73. Cf. the schema in footnote 2 above.

74. I. Dornbach, *Primatenkommunikation* (Frankfurt, 1975). On the relatively early differentiation of different types of speech action in the linguistic development of the child, see the pioneering dissertation of M. Miller, "Die Logik der frühen Sprachentwicklung" (Univ. of Frankfurt, 1975).

75. In a letter to me, G. Grewendorf cites the following counterexample: signing a contract, petition, etc. while simultaneously objectifying the corresponding illocutionary act. But only the following alternative seems possible: either the contract signing is legally carried out with the help of a performative utterance—in which case there is no objectification—or the nonverbal contract signing is accompanied by a statement: "S signs contract X"—in which case it is a question of two independent illocutionary acts carried out parallelly (where normally there is a division of roles: the statesman signs, the reporter reports the signing).

76. I. J. Cohen, "Do Illocutionary Forces Exist?," p. 587.

77. W. P. Alston, "Meaning and Use," in Rosenberg and Travis, *Readings,* p. 412: "I can find no cases in which sameness of meaning does not hang on sameness of illocutionary act."

78. For ontogenetic studies, a combination of a Piagetian theory of meaning for the cognitive schemata developed in connection with manipulated objects [cf. H. G. Furth, *Piaget and Knowledge* (Englewood Cliffs, N.J., 1969)] and a Meadian theory of meaning for the concepts developed in connection with interactions [cf. Arbeitsgruppe Bielefelder Soziologen, eds., *Alltagswissen, Interaktion und gesellschaftliche Wirklichkeit,* 2 vols. (Hamburg, 1973)] seems promising to me.

79. B. Richards argues against this in "Searle on Meaning and Speech Acts," *Foundations of Language* 7 (1971):536: "Austin argued that sentences such as Ra (I promise that I shall pay within one year) never *assert* anything that is either true or false, i.e., never assert propositions. Here we agree; but this in no way upsets the claim that Ra nevertheless *expresses* a proposition . . . viz. the proposition that Ra." Richards does not equate the propositional content of the speech action, Ra, with the propositional content of the dependent sentence: "I shall pay within one year," but with the content of the objectified speech action Ra, which must, however, then be embedded in a further speech action, Rv; for example, "I tell you, I promised him that I shall pay within one year." I regard the confusion of performative sentences with propositionally objectified sentences as a category mistake (which, incidentally, diminishes the value of Richards argument against Searle's principle of expressibility, in particular against his proposal to analyze the meaning of speech actions in standard form in terms of the meaning of the sentences used in the speech acts).

80. It follows from this proposal that each of the universal-pragmatic subtheories, that is, the theory of illocutionary acts as well as the theory of elementary sentences (and that of intentional expressions) can make its specific contribution to the theory of meaning. In Austin's choice of the terms *meaning* and *force*, there resonates still the descriptivist prejudice—a prejudice, I might add, that has been out of date since Wittgenstein at the latest, if not since Humboldt—according to which the theory of the elementary sentence, which is to clarify sense and reference, can claim a monopoly on the theory of meaning. (Naturally reference semantics also lives from this prejudice.)

81. Austin, *How to do Things with Words*, p. 132.

82. Ibid., pp. 147–148; Searle, *Speech Acts*, pp. 64 ff.

83. Austin, "Performative Utterances," p. 248.

84. Austin, *How to do Things with Words*, p. 144.

85. Ibid., pp. 145 ff. Cf. also "Performative-Constative," p. 31:

To begin with, it is clear that if we establish that a performative utterance is not unhappy, that is, that its author has performed his act happily and in all sincerity, that still does not suffice to set it beyond the reach of all criticism. It may always be criticized in a different dimension. Let us suppose that I say to you "I advise you to do it"; and let us allow that all the circumstances are appropriate, the conditions for success are fulfilled. In saying that, I actually do advise you to do it—it is not that I *state*, truely or falsely, *that* I advise you. It is, then, a performative utterance. There does still arise, all the same, a little question: was the advice good or bad? Agreed, I spoke in all sincerity, I believed that to do it would be in your interest; but was I right? Was my belief, in these circumstances, justified? Or again—though perhaps this matters less—was it in fact, or as things turned out, in your interest? There is confrontation of my utterance with the situation in, and the situation with respect to which, it was issued. I was fully justified perhaps, but was I right?

86. Austin, *How to do Things with Words*, pp. 144–145.

87. J. Habermas, "Vorbereitende Bemerkungen," pp. 111 ff.

88. Austin, "Performative Utterances," pp. 250–251.

89. Ibid., p. 251.

90. Austin, "Performative-Constative," pp. 31–32.

91. Searle, *Speech Acts*, p. 63.

92. On Wunderlich's analysis of advising (*Grundlagen*, pp. 349 ff.) the general context conditions would be as follows:

S makes it understood that (that is, S should give the advice only if these conditions obtain, and H should accordingly believe that they obtain):

1 S knows, believes, or assumes (depending on preceding communication) that
   a) H finds himself in an unpleasant situation Z;
   b) H wants or desires to reach some other, more pleasant situation $Z' \neq Z$;
   c) H does not know how Z' can be reached;
   d) H is in a position to do h.

2 S believes or assumes that
    e) H does not already want to do h;
    f) H can reach a more pleasant situation Z" (relative optimum) with h than with any alternative action h'.
3 S knows or believes, the following obligations are established for H: if one of the subconditions listed under (a) through (f) does not obtain (or more precisely, if H knows, believes, or assumes that it doesn't obtain), then H will make this understood to S in a conventional manner.

93. D. Holdcroft ignores this distinction, "Performatives and Statements," *Mind* 83 (1974):1–18, and thus comes to the false conclusion that only the speech actions we have called institutionally bound are subject to conventional regulations in the sense of the sentence: "A sentence type is a performative if and only if its literal and serious utterance can constitute the performance of an act which is done in accordance with a convention, which convention is not merely a grammatical or semantical one."
94. In Wunderlich's analysis of advising, his conditions B 4–6 make up the content of the obligations.
95. H. Delius, "Zum Wahrheitscharakter egologischer Aussagen," in Brockman and Hofer, eds., *Die Wirklichkeit des Unverständlichen* (The Hague, 1974), pp. 38–77.

Notes to "Moral Development and Ego Identity"

1. R. Döbert and G. Nunner-Winkler, "Konflikt- und Rückzugspotentiale in spätkapitalistischen Gesellschaften," in *Zeitschrift für Soziologie* (1973):301–325; R. Döbert and G. Nunner-Winkler, *Adoleszenzkrise und Identitätabildung* (Frankfurt, 1975).
2. J. Habermas, "Der Universalitätsanspruch der Hermeneutik," in *Kultur und Kritik* (Frankfurt, 1973), pp. 264–301. Parts of this have been translated in "On Systematically Distorted Communication," *Inquiry*, 13(1970):205–218.
3. H. Marcuse, "Das Veralten der Psychoanalyse," in *Kultur und Gesellschaft* 2 (Frankfurt, 1965), pp. 96–97; English translation, "The Obsolescence of the Freudian Concept of Man," in *Five Lectures* (Boston, 1970), pp. 44–61.
4. T. W. Adorno, "Zum Verhältnis von Soziologie und Psychologie," in *Sociologica* (Frankfurt, 1955), p. 43; English translation, "Sociology and Psychology," in *New Left Review*, 46 (1967):67–80 and 47 (1968): 79–97.
5. J. Habermas, *Legitimation Crisis* (Boston, 1975).
6. T. W. Adorno, *Negative Dialektik* (Frankfort, 1973), p. 294; English translation, *Negative Dialectic* (New York, 1973).

7. *Ego Psychology:* H. S. Sullivan, *Conceptions of Modern Psychiatry* (New York, 1940) and *The Interpersonal Theory of Psychiatry* (New York, 1953); E. H. Erikson, *Childhood and Society* (New York, 1963) and *Identity and the Life Cycle* (New York, 1959): N. Sanford, *Self and Society* (New York, 1966); D. J. de Levita, *Der Begriff der Identität* (Frankfurt, 1971); and G. and R. Blanck, "Toward a Psychoanalytic Developmental Psychology," in *Journal of the American Psychoanalytic Association* (1972):668–710.

Developmental Psychology: J. Piaget, *The Moral Judgement of the Child* (New York, 1965) and *Biology and Knowledge* (Chicago, 1971); H. Furth, *Piaget and Knowledge* (Englewood Cliffs, N.J., 1969); L. Kohlberg, "Stage and Sequence," in D. Goslin, ed., *Handbook of Socialization Theory and Research* (Chicago, 1969) and "From Is to Ought," in T. Mischel, ed., *Cognitive Development and Epistemology* (New York, 1971), pp. 151–236; J. H. Flavell, *The Development of Role-Taking and Communication Skills in Children* (New York, 1968); and H. Werner and B. Kaplan, *Symbol Formation* (New York, 1963).

Interactionism: C. H. Cooley, *Human Nature and the Social Order* (New York, 1902); G. H. Mead, *Mind, Self and Society* (Chicago, 1934); H. Gerth and C. W. Mills, *Character and Social Structure* (New York, 1953); T. Parsons and R. F. Bales, *Family Socialization and Interaction Process* (Glencoe, 1964), chap. 2, pp. 35–133; C. Gordon and K. J. Gergen, eds., *Self in Social Interaction* (New York, 1968); G. E. Swanson, "Mead and Freud, Their Relevance for Social Psychology," in J. G. Manis and B. N. Melzer, eds., *Symbolic Interaction* (Boston, 1967), pp. 25–45; L. Krappmann, *Soziologische Dimension de Identität* (Stuttgart, 1969); H. Dubiel, *Identität und Institution* (Bielefeld, 1973); and N. K. Denzin, "The Genesis of the Self in Early Childhood," in *The Sociological Quarterly* (1972):291–314.

8. J. Habermas, "Notizen zum Begriff der Rollenkompetenz,' in *Kultur und Kritik,* pp. 195–231.

9. "At the center of every psychological theory of development stands the concept of *developmental stages.* This has been worked out in its strongest and most precise form within the cognitivist tradition (Piaget, Kohlberg). These authors speak of stages of cognitive development only under the following conditions (J. H. Flavell, "An Analysis of Cognitive Developmental Sequences," in *Genetic Psychology Monographs* 86 (1972):279–350):(a) The cognitive schemata of the individual phases are *qualitatively* distinct from one another, and the individual elements of a phase-specific style of thought are so related to one another that they form a *structured whole.* Specific modes of behavior are not merely object-specific, externally stimulated responses; rather they can be interpreted as derivatives of a specific form of structuring the environment. (b) The phase-specific schemata are ordered in an *invariant and hierarchically structured sequence.* This means that no later phase can be attained with-

out passing through all those preceding it; further, that in later stages of development the elements of earlier phases are transformed (*aufgehoben*) and re-integrated at a higher level; and moreover that for the sequence as a whole a direction of development can be specified (growing independence from stimuli and greater objectivity). (c) These stages of development are *of psychological interest* above all for the following reason: from the fact that individuals always prefer problem-solutions corresponding to the highest level attained by them, and that schemata which spring from a superseded stage are in general avoided, we can infer that the development is not merely an externally constructed and imputed ordering schema, but corresponds to a psychologically and motivationally significant reality." R. Döbert and G. Nunner-Winkler, "Konflikt- und Rückzugspotentiale," p. 302.

10. J. Cumming and E. Cumming, *Ego and Milieu* (New York, 1967).

11. E. Turiel, "Conflict and Transition in Adolescent Moral Development," in *Child Development* (1974):14–29.

12. [Compare the discussion of "ego-demarcations" in the concluding section of "What is Universal Pragmatics?" *supra*.]

13. E. H. Erikson, *Identity and the Life Cycle* (New York, 1959), chap. 3, "The Problem of Ego Identity," pp. 101 ff.

14. J. Loevinger, "Origins of Conscience," unpubl. MS (Washington University, St. Louis, 1974).

15. J. Loevinger, "The Meaning and Measurement of Ego Development," in *American Psychologist* 21 (1966):195–206; J. Loevinger and R. Wessler, *Measuring Ego Development* (San Francisco, 1970); J. Loevinger, "Recent Research on Ego Development," unpubl. MS (Washington University, St. Louis, 1973); cf. also the dissertation by J. M. Broughton, *The Development of Natural Epistemology in Adolescence and Early Adulthood* (Harvard University, Cambridge, Mass., 1975).

16. T. Parsons, *The Social System* (London, 1951) and "Social Interaction," in the *International Encyclopaedia of Social Science*, vol. 7, pp. 429–441; J. Habermas, "Stichworte zur Theorie de Sozialisation," in *Kultur und Kritik*, pp. 118–194; and H. Joas, *Die gegenwärtige Lage der soziologischen Rollentheorie* (Frankfurt, 1973).

17. J. Habermas, "On Social Identity," in *Telos*, 19(1974):91–103.

18. A. W. Gouldner, "The Norm of Reciprocity," in *American Sociological Review* (1960):161–178; cf. also his *Enter Plato* (New York, 1965).

19. J. Sandler, "Zum Begriff des Über-Ichs," in *Psyche* (1964):721–743, 812–828; R. A. Spitz, *Genetic Field Theory of Ego Formation* (New York, 1959); E. Jacobson, *The Self and the Object World* (New York, 1964); and M. Mitscherlich, "Probleme der Idealisierung," in *Psyche* (1973):1106-1127.

20. A. Freud, *The Ego and the Mechanisms of Defense* (New York, 1946); G. E. Swanson, "Determinants of the Individual's Defenses against

Inner Conflict," in J. C. Glidewell, ed., *Parental Attitudes and Child Behavior* (Springfield, 1961), pp. 5 ff.; P. Madison, *Freud's Concept of Repression and Defense* (London, 1961). The attempt (in connection with the investigation of cognitive style) to find correlations between problemsolving and defense strategies (coping and defense mechanisms) is certainly interesting; cf. T. C. Kroeber, "The Coping Functions of the Ego-Mechanisms," in R. W. White, ed., *The Study Of Lives* (New York, 1963), pp. 178–200; N. Haan, "Tripartite Model of Egofunctioning," *Journal of Nervous and Mental Disease* (1969):14–30.

21. G. C. Gleser and D. Ihilebich, "An Objective Instrument for Measuring Defense-Mechanisms," in *Journal of Normal and Clinical Psychology* (1969):51–60; B. Neuendorff, *Geschlecht und Identität und die Struktur der Person-Umwelt-Interaktion,* dissertation (Berlin, 1976).

Notes to "Historical Materialism and the Development of Normative Structures"

1. A. Wellmer, *Critical Theory of Society* (New York, 1971), and "Communication and Emancipation: Reflections on the Linguistic Turn in Critical Theory," in John O'Neill, ed., *On Critical Theory* (New York, 1976), pp. 231–263; and J. Habermas, "Über das Subjekt der Geschichte," in Habermas, *Kultur und Kritik* (Frankfurt, 1973), pp. 389–398.

2. I. Fetscher, *Karl Marx und der Marxismus* (Munich, 1967); and O. Negt, "Marxismus als Legitimationswissenschaft," introduction to A. Deborin and N. Bucharin, *Kontroversen über dialektischen und mechanistischen Materialismus* (Frankfurt, 1969), pp. 7–50.

3. U. Oevermann, "Zur Theorie der individuellen Bildungsprozesse," unpubl. MS, Max-Planck-Institut für Bildungsforschung, Berlin, 1974.

4. H. Reichelt, *Zur logischen Struktur des Kapitalbegriffs bei K. Marx* (Frankfurt, 1970).

5. H. J. Sandkühler and R. de la Vega, eds., *Marxismus and Ethik* (Frankfurt, 1974).

6. K. O. Apel, "Sprechakttheorie und die Begründung der Ethik," in K. O. Apel, ed., *Sprachpragmatik und Philosophie* (Frankfurt, 1976).

7. J. Habermas, "What is Universal Pragmatics?," *supra.*

8. J. Habermas, *Legitimation Crisis* (Boston, 1975), pp. 8 ff.

9. J. Piaget, *The Moral Judgment of the Child* (New York, 1965); and L. Kohlberg, "Stage and Sequence," in D. Goslin, ed., *Handbook of Socialization Theory and Research* (Chicago, 1969), and "From is to Ought," in T. Mischel, ed., *Cognitive Development and Epistemology* (New York, 1971), pp. 151–236.

10. [The first of these three "domains"—law and morality—is dealt with in the next essay. What follows is a brief sketch of the types of "homologies" conjectured for the other two domains, world-views and group identities.]

11. This is not a question of linear development; there are phases of regression in the transition from one stage to the next. Cf., e.g., D. Elkind, "Egocentrism in Adolescence," in *Child Development* 38 (1967):1025–1034.

12. J. Piaget, *The Child's Conception of Time* (New York, 1970), and *The Child's Conception of Physical Causality* (London, 1966).

13. R. Döbert, "Modern Religion and the Relevance of Religious Movements," unpubl. MS, Max-Planck-Institut zur Erforschung der Lebensbedingungen der wissenschaftlichtechnischen Welt, Starnberg, 1975.

14. B. Neuendorff, "Geschlechtliche Identität. Zur Strukturierung der Person-Unwelt-Interaktion," Diss., Free University of Berlin, 1976.

15. J. Habermas, "Notizen zum Begriff der Rollenkompetenz," in *Kultur und Kritik*, pp. 195–231.

16. E. Goffman, *Stigma* (Englewood Cliffs, N.J., 1963).

17. G. H. Mead, *Mind, Self, and Society From the Standpoint of a Social Behaviorist*, C. W. Morris, ed. (Chicago, 1934), and *The Social Psychology of George Herbert Mead*, A. Strauss, ed. (Chicago, 1956).

18. M. Looser, "Personalpronomen und Subjektivität," in A. Leist, *Materialistische Sprachtheorie* (Kronberg, 1975); and cf. the bibliography provided there.

19. R. Döbert, G. Nunner-Winkler, *Adoleszenzkrise und Identitätsbildung* (Frankfurt, 1975); and H. Stierlein, *Eltern und Kinder im Prozess der Ablösung* (Frankfurt, 1975).

20. Cf. the schema developed by U. Oevermann, reproduced in *Kultur und Kritik*, p. 231. On the critique of conventional role theory, see L. Krappmann, *Dimensionen der Identität* (Stuttgart, 1969); H. Joas, *Zur gegenwärtigen Lage der soziologischen Rollentheorie* (Frankfurt, 1973); H. Dubiel, *Identität und Institution* (Gütersloh, 1973); and D. Geulen, "Das vergesellschaftete Subjekt," 2 vols., Diss., Free University of Berlin, 1975.

21. [Habermas may be referring here to "Notizen zum Begriff der Rollenkompetenz," pp. 222 ff.]

22. T. Luckmann, "On the Boundaries of the Social World," in M. Natanson, ed., *Phenomenology and Social Reality* (The Hague, 1970), pp. 73–100.

23. V. Lanternari, *The Religions of the Oppressed; a Study of Modern Messianic Cults* (London, 1963).

24. G. Devereux, *Normal und Anormal* (Frankfurt, 1974).

25. Cf. J. Habermas, "Überlegungen zum evolutionären Stellenwert des modernen Rechts," in Habermas, *Zur Rekonstruktion des Historischen Materialismus* (Frankfurt, 1976), pp. 260–267.

26. J. Habermas, *Strukturwandel der Öffentlichkeit* (Neuwied, 1962), parts I and II.

27. Marx analyzed this double identity—with reference naturally to Rousseau—in "On the Jewish Question," *Writings of the Young Marx*

on Philosophy and Society, L. D. Easton and K. H. Guddat, eds. and trans. (New York, 1967), pp. 216–248.

28. W. Wette, "Bundeswehr ohne Feindbilder?," in *Friedensanalysen,* 1 (Frankfurt, 1975), pp. 96–114.

29. Cf. e.g., B. R. Dulong, *La question Bretonne* (Paris, 1975).

30. In discussion, Klaus Eder advanced the thesis that there could be collective identities corresponding to personal identities only at the stage of conventional (role) identity. Postconventional *ego* identity would have to do without support from a collective identity. The fictions of a cosmopolitan state of affairs, a socialist order of society, an association of free producers, and so on, would then merely signify stages in the dissolution of collective identity as such. Kant, for example, represented the intelligible world as a "universal kingdom of ends-in-themselves." He saw that "the concept of the ethical community is always related to the idea of a totality of *all* men; it is distinguished therein from the concept of the political [community]." The kingdom of rational beings is an ideal that can never be empirically fulfilled by the just order of a cosmopolitan state of affairs. Nevertheless, such identity projections illustrate the conditions of a universalistically regulated domain of communicative action, against which the provisionally constructed, collective identities of particular reference groups can be relativized and rendered fluid. From this perspective the question of whether complex societies can construct a rational identity would have to be answered as follows: a collective identity becomes superfluous as soon as the mass of the members of society are sociostructurally forced to lay aside their role identities, however generalized, and to develop ego identities. The idea of an identity become reflective, to be collectively established in the future, would now be the last illusory husk before collective identities as such could be given up and replaced by the permanent variation of all reference systems. Such a state of affairs also bears utopian features; for in it all wars—as organized efforts of collectives that demand from their members a willingness to die—would be thinkable only as regressive states of emergency but no longer as something institutionally expected to happen.

31. J. Habermas, *Knowledge and Human Interests* (Boston, 1971), and *Toward a Rational Society* (Boston, 1970). I am grateful to T. McCarthy for his contributions to the analysis of the concepts of instrumental, strategic, and communicative action [in *The Critical Theory of Jürgen Habermas* (Cambridge, Mass., 1978), section 1.2, "Labor and Interaction: The Critique of Instrumental Reason"]. Cf. also J. Keane, "Work and Interaction in Habermas," in *Arena* 38 (1975):51–68.

32. On what follows, cf. also A. Wellmer, "Communication and Emancipation . . . ."

33. On the concept of systematically distorted communication, cf. J. Habermas, "Der Universalitätsanspruch der Hermeneutik," in *Kultur und Kritik,* pp. 263–301. [Some of the same ground is covered in "Sum-

mation and Response," *Continuum* 8 (1970):123–133, and "On systematically distorted Communication," *Inquiry* 13 (1970):205–218.]

34. In this context I am primarily emphasizing the difference between
rationalization processes attaching to the different aspects of action. Marx
attempted to conceive the *unity* of these rationalization processes, drawing on the Hegelian dialectic to characterize the relation of the individual
to society in the precapitalist, capitalist, and postcapitalist periods. C.
Gould makes an interesting attempt at reconstruction in *Marx's Social
Ontology* (Cambridge, Mass., 1978).

35. J. H. Flavell, "An Analysis of Cognitive Developmental Sequences,"
Genetic Psychology Monographs 86 (1972):279–350.

36. M. Godelier, *Rationalité et irrationalité en économie* (Paris, 1966),
and *Perspectives in Marxist Anthropology* (Cambridge, 1976). Godelier's
work is, of course, based on that of Levi-Strauss.

37. L. Goldmann, *Structures Mentales et Création Culturelle* (Paris,
1970), *Marxisme et Sciences Humaines* (Paris, 1970), and *La Création
Culturelle dans la Société Moderne* (Paris, 1971).

38. A. Touraine, *Production de la Société* (Paris, 1972), and *Pour la
Sociologie* (Paris, 1974).

39. M. Vester, *Die Entstehung des Proletariats als Lernprozess* (Frankfurt, 1970); and O. Negt and A. Kluge, *Öffentlichkeit und Erfahrung*
(Frankfurt, 1972).

40. C. Offe, *Strukturprobleme des kapitalistischen Staates* (Frankfurt,
1972); cf. also M. Jänicke, *Politische Systemkrisen* (Köln, 1973); and
W. D. Narr and C. Offe, eds., *Wohlfahrtsstaat und Massenloyalität*
(Köln, 1975).

41. C. Offe, *Berufsbildungsreform* (Frankfurt, 1975).

42. T. W. Adorno, "Kultur und Verwaltung," in *Soziologische Schrift-
en 1, Gesammelte Schriften,* vol. 8, pp. 122–146, and "Spatkapitalismus
oder Industriegesellachaft," ibid., pp. 354–372.

43. M. Greiffenhagen, ed., *Demokratisierung in Staat und Gesellschaft*
(Munich, 1973); and H. von Hentig, *Die Wiederherstellung der Politik*
(Stuttgart, 1973).

44. [In the original there follow two paragraphs in which Habermas
outlines the contents of the book, acknowledges the contributions of R.
Döbert and K. Eder to the ideas developed therein, and again emphasizes their programmatic character.]

Notes to "Toward a Reconstruction of Historical Materialism"

1. In the first part of *The German Ideology* and in the preface to *A
Contribution to the Critique of Political Economy.*

2. On the relationship of the assessments of historical materialism by
Marx and Engels, cf. L. Krader, *Ethnologie und Anthropologie bei Marx*
(München, 1973).

3. J. Stalin, *Dialectical and Historical Materialism* (New York, 1940).

4. I. S. Kon, *Die Geschichtsphilosophie des 20. Jahrhunderts*, vol. 2 (Berlin, 1966); E. M. Zukov, "Über die Periodisierung der Weltgeschichte," *Sowetwissenschaft* (1961, no. 3):241–254; E. Engelberg, "Fragen der Evolution und der Revolution in der Weltgeschichte," *Zeitschrift für Geschichtswissenschaft*, 13 (1965):9–18; E. Hoffman, "Zwei aktuelle Probleme der geschichtlichen Entwicklungsfolge fortschreitender Gesellschaftsformationen," *Zeitschrift für Geschichtswissenschaft*, 16(1968): 1265–1281; G. Lewin, "Zur Diskussion über die marxistische Lehre von den Gessellschaftsformationen," *Mitteilungen des Instituts für Orientforschung* (1969):137–151; and E. Engelberg, ed., *Probleme der marxistischen Geschichtswissenschaft* (Köln, 1972).

5. Marx and Engels, *The German Ideology*, in L. Easton and K. Guddat, eds., *Writings of the Young Marx on Philosophy and Society* (New York, 1967), p. 409.

6. On the delimitation of action types, cf. J. Habermas, *Toward a Rational Society* (Boston, 1970), pp. 91 ff.

7. Marx and Engels, *German Ideology*, p. 421.

8. Ibid., p. 409.

9. Ibid., p. 402.

10. B. Rensch, *Homo Sapiens: From Man to Demi-God* (New York, 1972); E. Morin, *Das Rätsel des Humanen* (München, 1974).

11. C. F. Hockett and R. Ascher, "The Human Revolution," *Current Anthropology* (Feb. 1964):135–147; G. W. Hewes, "Primate Communication and the Gestural Origin of Language," *Current Anthropology* (Feb. 1973):5–29.

12. On incest barriers among vertebrates, cf. N. Bischoff, "The Biological Foundations of the Incest-taboo," *Social Science Information* 6 (1972):7–36. Ethological investigations do not take into account that it is the incest barrier between father and daughter that first clears the culturally innovative way to the family structure. Cf. Meyer Fortes, "Kinship and the Social Order," *Current Anthropology* (April 1972):285–296.

13. E. W. Count, *Being and Becoming: Essays on the Biogram* (New York, 1973).

14. J. Habermas, "Entwicklung der Interaktionskompetenz," unpubl. MS (Starnberg, 1974).

15. Morin, *Das Rätsel des Humanen*, pp. 115 ff. On the ontogenesis of time consciousness, cf. J. Piaget, *The Child's Conception of Time* (New York, 1970).

16. D. Claessens, *Instinkt, Psyche, Geltung* (Opladen, 1967). Durkheim has already investigated the obligatory character of action norms, which to begin with generate their own power of sanction, under the aspect of the binding of feeling ambivalence; cf. E. Durkheim, *Sociology and Philosophy* (New York, 1974): "Furthermore there is another con-

cept that exhibits the same duality, namely that of the holy. The holy object instills in us, if not fear, then certainly respect, which keeps us at a distance from it. At the same time it is an object of love and desire; we aspire to get closer to it, we strive toward it. Thus we have to do here with a double feeling." Cf. also A. Gehlen's theses on "indeterminate obligations" in *Urmensch und Spätkultur* (Bonn, 1956), pp. 154 ff.

17. On the concepts of "internal" and "external" nature, cf. J. Habermas, *Knowledge and Human Interests* (Boston, 1971), and *Legitimation Crisis* (Boston, 1975), pp. 8 ff.

18. Stalin, *Dialectical and Historical Materialism*.

19. J. Pecirka, "Von der asiatischen Produktionsweise zu einer marxistischen Analyse der frühen Klassengesellschaften," *Eirene* 6 (Prague, 1967), pp. 141–174; and L. V. Danilova, "Controversial Problems of the Theory of Precapitalist Societies," *Soviet Anthropology and Archeology*, 9 (Spring 1971): 269–327.

20. M. Godelier, *Perspectives in Marxist Anthropology* (Cambridge, 1976).

21. Recently, O. Marquardt, *Schwierigkeiten mit der Geschichtsphilosophie* (Frankfurt, 1973).

22. Cf. the preceding essay.

23. In an unpublished manuscript on the theory of evolution, Niklas Luhmann expresses doubts about the applicability of the concept of motion in this connection.

24. Luhmann points this out in the manscript mentioned in n. 23.

25. Cf. my critique of Luhmann in J. Habermas, N. Luhmann, *Theorie der Gesellschaft* (Frankfurt, 1971), pp. 150 ff; cf. also R. Döbert, *Systemtheorie und die Entwicklung religiöser Deutungssysteme* (Frankfurt, 1973), pp. 66 ff.

26. For example, H. Gericke, in "Zur Dialektik von Produktivkraft und Produktionsverhältnis im Feudalismus," *Zeitschrift für Geschichtswissenschaft*, 16(1966):914–932, distinguishes the "increasingly higher degree of mastery of nature" from the "increasingly maturer forms of corporate social life": "The most important criteria and the decisive factors in historical progress are improvement of productive forces, especially the increase in conscious, goal-directed, success-oriented activity of immediate producers, as well as altered productive relations, which permit an ever increasing number of people to participate competently and actively in economic, social, political, and cultural processes" (pp. 918–919).

27. K. Marx, *A Contribution to the Critique of Political Economy*, M. Dobb, ed. (New York, 1970), Preface, pp. 20–21.

28. K. Kautsky, *Die materialistische Geschichtsauffassung*, 2 vols. (Berlin, 1927), vol. 1, pp. 817–818.

29. Cf. A. Touraine, *La Société post-industrielle* (Paris, 1969); and D. Bell, *The Coming of Postindustrial Society* (New York, 1973).

30. Godelier, *Perspectives in Marxist Anthropology.*

31. Marx, *A Contribution,* Preface, p. 21.

32. Stalin, *Dialectical and Historical Materialism.*

33. Godelier, *Perspectives in Marxist Anthropology.*

34. Marx and Engels, *Manifesto of the Communist Party,* in L. Feuer, ed., *Marx and Engels: Basic Writings on Politics and Philosophy* (New York, 1959), p. 19.

35. I. Sellnow, "Die Auflösung der Urgemeinschaftsordnung," in K. Eder, ed., *Die Entstehung von Klassengesellschaften* (Frankfurt, 1973), pp. 69–112.

36. S. Moscovici, *The Human History of Nature* (New York, 1977).

37. Marx, *A Contribution,* Preface, p. 21.

38. A. Gehlen, "Anthropologische Ansicht der Technik," in Gehlen, *Technik im technischen Zeitalter* (Düsseldorf, 1965); and cf. my remarks in *Toward a Rational Society,* pp. 87 ff.

39. J. Piaget, *The Principles of Genetic Epistemology* (New York, 1972).

40. E. C. Welskopf, "Schauplatzwechsel und Pulsation des Fortschritts," in E. Schulin, ed., *Universalgeschichte* (Köln, 1974), pp. 122–123.

41. Hoffmann, "Zwei aktuelle Probleme."

42. V. G. Childe, *What Happened in History* (London, 1964), pp. 55 ff; C. M. Cipolla, "Die zwei Revolutionen," in *Universalgeschichte,* pp. 87–95.

43. Danilova, "Controversial Problems," pp. 282–283.

44. E. R. Service, *Primitive Social Organization* (New York, 1962).

45. K. Eder, *Die Entstehung staatlich organisierter Gesellschaften* (Frankfurt, 1976).

46. Pecirka, "Von der asiatischen Produktionsweise."

47. R. Gunther, "Herausbildung und Systemcharakter der vorkapitalistischen Gesellschaftsformationen," *Zeitschrift für Geschichtswissenschaft,* 16 (1968):1204–1211.

48. Gericke, "Zur Dialektik."

49. These phenomena stimulated Karl Jaspers to construct his notion of an "axial period"; cf. his *The Origin and Goal of History* (New York, 1976).

50. F. Tökei, *Zur Frage der asiatischen Produktionsweise* (Neuwied, 1965).

51. M. Finley, "Between Slavery and Freedom," *Comparative Studies in Society and History,* 6(April, 1964).

52. J. Habermas, *Legitimation Crisis,* pp. 17 ff.; K. Eder, "Komplexität, Evolution und Geschichte," in *Theorie der Gesellschaft, Supplement I* (Frankfurt, 1973).

53. L. Kohlberg, "Stage and Sequence," in D. Goslin ed., *Handbook*

of Socialization Theory and Research (Chicago, 1969), and "From Is to Ought," in T. Mischel, ed., *Cognitive Development and Epistemology* (New York, 1971), pp. 151–236.

54. Eder, *Die Entstehung staatlich organisierter Gesellschaften.*

55. L. Krader, *Formation of the State* (New York, 1968).

56. The most important representatives of this theory are F. Ratzel, P. W. Schmidt, F. Oppenheimer, and A. Rüstow.

57. W. E. Mühlmann, "Herrschaft und Staat," in *Rassen, Ethnien, Kulturen* (Neuwied, 1964), pp. 248–296.

58. This view, first developed by Marx and Engels in *The German Ideology*, has had many adherents; V. G. Childe is a good representative, originally in *Old World Prehistory* (London, 1938).

59. G. E. Lenski, *Power and Privilege* (New York, 1966). Earlier I too defended this view; cf. *Toward a Rational Society*, p. 94, and *Theorie der Gesellschaft*, pp. 153–175.

60. R. L. Carneiro, "A Theory of the Origin of the State," *Science* 169(1970):733–738.

61. K. A. Wittfogel, *History of Chinese Society* (Philadelphia, 1946), and *Oriental Despotism: A Comparative Study of Total Power* (New Haven, 1957).

62. R. Coulborn, "Structure and Process in the Rise and Fall of Civilized Societies," *Comparative Studies in History and Society* 8 (1965–66); Carneiro, "A Theory of the Origin of the State."

63. I am drawing on a sketch presented by Klaus Eder at the *16. Deutschen Soziologentag* in Kassel, 1974.

64. Ibid., p. 14.

65. Ibid., p. 15.

66. Ibid.

67. "The deep-seated contradiction was that the mastery of nature and the self-realization of man sometimes had to come into opposition, since the former process required for its increasing efficacy servitude as a means of realizing the organization and mobility (of labor power), while the latter had freedom as its goal and basis. Indeed in the final analysis the mastery of nature makes sense only if the self-realization of man, the humanizing of relations among men, succeeds." Welskopf, "Schauplatzwechsel," p. 131.

68. K. Popper, The *Poverty of Historicism* (London, 1966).

69. C. H. Waddington, *The Ethical Animal* (Chicago, 1960).

70. Cf. W. Leppenies, H. H. Ritter, eds., *Orte des wilden Denkens* (Frankfurt, 1970).

71. Piaget also singles this out as the moment that unites the different brands of structuralism; cf. J. Piaget, *Structuralism* (New York, 1970).

72. C. Lévi-Strauss, *The Savage Mind* (Chicago, 1968); M. Godelier, "Mythe et Histoire," *Annales, Economies, Sociétés, Civilisations.*

73. L. Kohlberg, "Stage and Sequence" and "From Is to Ought."

74. S. Goeppert, H. C. Goeppert, *Sprache und Psychoanalyse* (Heidelberg, 1973); R. Döbert, "Zur Logik des Ubergangs von archaischen zu hochkulturellen Religionssytemen," in K. Eder, ed., *Die Entstehung von Klassengesellschaften,* pp. 330–363; B. Schlieben-Lange, *Linguistische Pragmatik* (Stuttgart, 1975).

75. V. G. Childe, *What Happened in History* (New York, 1946); L. A. White, *The Science of Culture* (New York, 1949).

76. J. H. Steward, *Theory and Culture Change* (Urbana, 1955).

77. T. Parsons, *Societies: Evolutionary and Comparative Perspectives* (Englewood Cliffs, N.J., 1966); G. Lenski, *Human Societies* (New York, 1930); cf. the critical remarks of P. J. Utz, "Evolutionism Revisited," *Comparative Studies in Society and History,* 15 (1973):227–240; N. Luhmann, *Zweckbegriff und Systemrationalität* (Frankfurt, 1974).

78. E. S. Dunn, *Economic and Social Development* (Baltimore, 1971), pp. 80 ff.

79. Ibid.

80. Ibid., pp. 160 ff.

81. In an unpublished manuscript on the theory of social evolution.

82. Dunn, *Economic and Social Development,* pp. 97 ff.

83. H. W. Nissen, "Phylogenetic Comparison," in S. S. Stevens, ed., *Handbook of Experimental Psychology* (New York, 1951), pp. 34 ff.

84. N. Luhmann, "Einführende Bemerkungen zu einer Theorie symbolisch generalisierter Kommunikationsmedien," *Zeitschrift fur Soziologie* (June 1974):236–255.

85. I owe this point to a conversation with Klaus Eder.

86. J. Huxley, *Evolution: The Modern Synthesis* (New York, 1941), and *Touchstone for Ethics* (New York, 1942); T. Dobzhansky, *The Biological Basis of Human Freedom* (New York, 1956); and D. D. Raphael, "Darwinism and Ethics," in S. A. Barnett, ed., *A Century of Darwin* (Cambridge, 1958).

87. Waddington, *The Ethical Animal,* p. 59.

88. Ibid.

89. K.-O. Apel, "Das Apriori der Kommunikationsgemeinschaft und die Grundlagen der Ethik," in Apel, *Transformation der Philosophie* (Frankfurt, 1973), vol, 2, pp. 358–436, and J. Habermas, "What is Universal Pragmatics?," *supra.*

90. Cf. J. Habermas, "Zwei Bemerkungen zum praktischen Diskurs," in Habermas, *Zur Rekonstruktion des Historischen Materialismus* (Frankfurt, 1976), pp. 338–346.

Notes to "Legitimation Problems in the Modern State"

1. In a certain way the kinship system itself has legitimating power; the status of the family to which one belongs decides which claims one may put forward. The concept of "legitimate heirs" in Roman law trans-

ports this meaning into civil law. Legitimation in the sense of private entitlement presupposes, however, a legitimate order.

2. K. Eder, *Die Entstehung staatlich organisierter Klassengesellschaften* (Frankfurt, 1976).

3. S. Rokkan, "Die vergleichende Analyse der Staaten- und Nationenbildung," in W. Zapf, ed., *Theorien des sozialen Wandels* (Köln, 1969), pp. 228–252.

4. For this reason, the Aristotelian concept of the *polis* is less a constitutional concept than a concept of identity. Cf. J. Ritter, "Politik und Ethik in der praktischen Philosophie des Aristoteles," in Ritter, *Metaphysik und Politik* (Frankfurt, 1969), pp. 106–132.

5. N. Luhmann, "Die Weltgesellschaft," *Archiv für Rechts- und Sozialphilosophie* (1971):1–33; see also my critical remarks in "Konnen komplexe Gesellschaften eine vernünftige Identität ausbilden?," in Habermas, *Zur Rekonstruktion des Historischen Materialismus* (Frankfurt, 1976), pp. 92–126; some of these remarks are translated in "On Social Identity," *Telos* 19(1974):91–103.

6. Cf. the argument of P. Kielmannsegg, "Legitimität als analytische Kategorie," *Politische Vierteljahresschrift* 12 (1971):367–401, esp. pp. 391 ff.

7. T. Würtenberger, *Die Legitimität staatlicher Herrschaft* (Berlin, 1973).

8. C. Meier, "Die Entstehung des Begriffs 'Demokratie,'" *Politische Vierteljahresschrift* 10(1969):535–575.

9. V. Lanternari, *The Religions of the Oppressed: A Study of Modern Messianic Cults* (London, 1963), p. 316. "When the pressure of the white man makes itself felt from within a society, the natives reach for the Bible, which they had refused to accept from the missionaries during centuries of evangelism. This "self-Christianization" of many native groups came about when the whites, having forced their way into the native environment, created conditions similar to those which fostered the early growth of Christianity. As it was for the first Christians of the Middle East and of ancient Rome, so it was for the native peoples of Africa, Asia, Oceania, and the Americas: pressure and opposition came upon them simultaneously from two sides, the militant hierarchy of the church and the authoritarian power of the state."

10. The proposed laws—land reform, shortening of military service, civil rights for allies—throw light on the background of class confrontations between aristocratic owners of large landed estates and peasants. The attempt to set up a democracy on the Greek model, to withdraw as many matters as possible from the responsibility of the Senate and transfer them to the popular assembly, to alter the composition of the courts made up of senators, shows that we have to do with a legitimacy conflict. That Octavius was unconstitutionally removed from his office of tribune, that Tiberius allowed himself to be illegally put up as a candi-

date for a second time, that the Senate did not legally pursue the murder of Tiberius in the streets—all this speaks for a radical decline in the legitimacy of the existing order.

11. E. Werner, M. Erbstösser, *Ideologische Probleme des mittelalterlichen Plebejertums* (Berlin, 1960).

12. M. Becker, "Florentine Politics and the Diffusion of Heresy in the Trecento," *Spectaculum* 34(1959):67–75.

13. N. Cohen, *The Pursuit of the Millenium* (London, 1957); J. B. Russel, *Religious Dissent in the Middle Ages* (New York, 1971).

14. K. Jaspers, *The Great Philosophers,* H. Arendt, ed., (New York, 1962).

15. J. Rawls, *A Theory of Justice* (Cambridge, Mass., 1971).

16. K.-O. Apel, "Das Apriori der Kommunikationsgemeinschaft und die Grundlagen der Ethik," in *Transformation der Philosophie,* vol. 2 (Frankfurt, 1973), pp. 358–436.

17. R. Döbert, "Zur Logik des Übergangs von archaischen zu hochkulturellen Religionssystemen," in K. Eder, ed., *Die Entstehung von Klassengesellschaften* (Frankfurt, 1973), pp. 330–363.

18. Rousseau, *The Social Contract,* Book I, Chapter 6.

19. H. Grebing, "Volksrepräsentation und identitäre Demokratie," *Politische Vierteljahresschrift* (1972):162–180; J. Fijalkowski, "Bemerkungen zu Sinn und Grenzen der Rätediskussion," in M. Greiffenhagen, *Demokratisierung in Staat und Gesellschaft* (München, 1973), pp. 124–139; F. Scharpf, "Demokratie als Partizipation," in ibid., pp. 117–124; and H. von Hentig, *Die Wiederherstellung der Politik* (Stuttgart, 1973).

20. P. Bachrach, *The Theory of Democratic Elitism: A Critique* (Boston, 1967); C. Pateman, *Participation and Democratic Theory* (Cambridge, 1970); and Q. Skinner, "The Empirical Theorists of Democracy and Their Critics: A Plague on Both Their Houses," *Political Theory* 1(1973):287–306.

21. Von Kielmannsegg, "Legitimität als analytische Kategorie," p. 381.

22. N. Luhmann, *Legitimation durch Verfahren* (Neuwied, 1969).

23. J. Habermas, *Legitimation Crisis* (Boston, 1975), pp. 142–143.

24. L. Basso, "Gesellschaft und Staat in der Marxschen Theorie," in Basso, *Gesellschaftsformation und Staatsform* (Frankfurt, 1975), pp. 10–46.

25. C. Offe, *Bildungsreform* (Frankfurt, 1975), pp. 24–25.

26. Cf. my concept of a "system crisis" in *Legitimation Crisis,* pp. 24–31.

27. C. Tilly, "Reflections on the History of European State-Making," in Tilly, ed., *The Formation of National States in Western Europe* (Princeton, 1975), pp. 3–83.

28. I. Wallerstein, *The Modern World-System* (New York, 1974).

29. S. E. Finer, "State- and Nation-Building in Europe: The Role of the Military," in Tilly, ed., *The Formation of National States,* pp. 84–163.

30. T. Würtenberger, *Die Legitimation staatlicher Herrschaft* (Berlin, 1973).

31. D. Sternberger, "Legitimacy," *International Encyclopedia of Social Science,* vol. 9, pp. 244–248.

32. This expression is used by C. J. Friedrich, "Die Legitimität in politischer Perspektive," *Politische Vierteljahresschrift* (1960).

33. C. B. McPherson, *The Political Theory of Possessive Individualism: Hobbes to Locke* (Oxford, 1962); W. Euchner, *Egoismus und Gemeinwohl* (Frankfurt, 1973); and H. Neuendorff, *Der Begriff des Interesses* (Frankfurt, 1973).

34. H. U. Wehler, *Geschichte des Deutschen Kaiserreichs* (Göttingen, 1974).

35. C. Tilly, "Food Supply and Public Order in Modern Europe," in *The Formation of National States,* pp. 380–456.

36. C. Offe and V. Ronge, "Thesen zur Begründung des Konzepts des kapitalistischen Staates," unpubl. MS (Starnberg, 1975); and C. Offe, *Strukturprobleme des kapitalistischen Staates* (Frankfurt, 1972).

37. S. Skarpelis-Sperk et al., "Ein biedermeierlicher Weg zum Sozialismus," *Spiegel, no. 9, 1975.*

38. J. Habermas, *Toward a Rational Society* (Boston, 1970); C. Koch and D. Senghass, *Texte zur Technokratiediskussion* (Frankfurt, 1970); and J. Habermas, *Legitimation Crisis,* pp. 130–142.

39. O. Massing, "Restriktive sozio-ökonomische Bedingungen parlamentarischer Reformstrategien," in *Politische Soziologie* (Frankfurt, 1974), pp. 123–138.

40. B. Guggenberger, "Herrschaftslegitimierung und Staatskrise," in Greven, Guggenberger, and Strasser, *Krise des Staates* (Neuwied, 1975), pp. 9–10; J. O'Connor, *The Fiscal Crisis of the State* (New York, 1973).

41. Fröbel, Heinrichs, Kreye, and Sunkel, "Internationalisierung von Arbeit und Kapital," *Leviathan* 1(1973):429–454.

42. C. Offe and V. Ronge, "Thesen."

43. R. Miliband, *The State in Capitalist Society* (New York, 1969).

44. R. M. Merelman, "Learning and Legitimacy," *American Political Science Review* 60(1966), p. 548.

45. H. Busshoff, *Systemtheorie als Theorie der Politik* (München, 1975).

46. K. D. Opp, "Einige Bedingungen für die Befolgung von Gesetzen," in K. Lüderssen and F. Sack, eds., *Abweichendes Verhalten I* (Frankfurt, 1975), pp. 214–243.

47. J. H. Schaar, "Legitimacy in the Modern State," in P. Green and S. Levinson, eds., *Power and Community* (New York, 1970, pp. 277–327; R. Spaemann, "Die Utopie der Herrschaftsfreiheit," in M. Riedel, ed., *Rehabilitierung der praktischen Philosophie II* (Freiburg, 1974), pp. 211–234.

48. J. Ritter, "Naturrecht bei Aristoteles," in *Metaphysik und Politik,* pp. 133–182, here p. 135.

49. Ibid., p. 169.

50. H. G. Gadamer, "Uber die Möglichkeit einer philosophischen Ethik," in Gadamer, *Kleine Schriften I*, pp. 179–191, here pp. 187–189; W. Hennis, *Politik und praktische Philosophie* (Neuwied, 1963); cf. also H. Kuhn, "Aristoteles und die Methode der politischen Wissenschaft," in M. Riedel, *Rehabilitierung*, pp. 161–190.

51. H. Pitkin, *Wittgenstein and Justice* (Berkeley, Calif., 1972), pp. 169–192.

52. Ibid., p. 272.

53. M. Oakeshott, *Rationalism in Politics and Other Essays* (New York, 1962).

54. I am taking up a suggestion of V. Held, *The Public Interest* (New York, 1970). Held in turn relies on H. L. A. Hart, *The Concept of Law* (Oxford, 1961).

55. Rawls, *A Theory of Justice:* P. Lorenzen, *Normative Logic and Ethics* (Mannheim, 1969); F. Kambartel, "Wie ist praktische Philosophie konstruktiv möglich?," in Kambartel, ed., *Praktische Philosophie und konstruktive Wissenschaftstheorie* (Frankfurt, 1974), pp. 6–33. K.-O. Apel, "Sprechakttheorie und transzendentale Sprachpragmatik: Zur Frage ethischer Normen," in Apel, ed., *Sprachpragmatik und Philosophie* (Frankfurt, 1976), pp. 10–173. On the situation of the discussion in the German-speaking world, cf. M. Riedel, ed., *Rehabilitierung der praktischen Philsophie, I–II* (Freiburg, 1972, 1974); see also R. Bubner, "Eine Renaissance der praktischen Philosophie," *Philosophische Rundschau* 22 (1975):1–34.

56. Cf. the two preceding essays.

INDEX

Action systems, 155–158
Action-theoretic framework, 78, 82
Adler, M., 143
Adorno, T., 71–72, 120, 143
Alston, W. P., 45
Althusser, L., 167
Apel, K. O · theory of speech actions in, 1–2, 5; transcendental analysis in, 22–24, 184, 205
Arendt, H., 201
Aristotle, 191; politics of, viii, 201–202
Austin, J. L., 7, 26, 216n, 217n; doctrine of infelicities in, 60; illocutionary act in, 34, 44–50; validity claims of speech in, 50–53, 55, 57

Bar-Hillel, Y., 209n
Bever, B. G., 20, 212n
Botha, R. P., 211n

Capitalist economy, 105, 144, 152, 188, 189; development of, 130, 139, 146, 150, 152; legitimation problems in, 165, 194–198; organizational principle of, 114, 123–124, 126–128, 190
Carnap, R., 5, 13
Chomsky, N., 96; essentialist claims of, 16, 19–20; reconstructive analysis in, 14, 16–17
Class societies, emergence of, 158, 161–163; legitimation problems in, 163; structure of, 181–182, 192–193

Cognitive competence, 14, 82–83, 106, 167–168
Cognitive development, xx, 20, 104; and ego-development, 73, 75, 78, 91, 100–103, 111, 121; and history of technology, 149, 169; potentials of, 140, 145, 147; psychology of, 25, 73, 149, 160, 171–172, 205
Collective identity, 108, 110–116, 120, 161, 223n; and legitimation problems, 179–180, 182; and modern state structure, 190–194
Communication, modes of, 50–58, 63
Communicative action, 1, 34, 38, 208n; and action structures, 41, 117–120, 145; and ego-development, 78, 82–86, 91; and historical materialism, 95–99; reciprocity in, 88–90, 107, 109; rules of, 132, 137–138, 148; validity claims of, 3–4, 35, 57–59, 66, 177, 200
Communicative competence, xii, xviii–xix, xxi, 26–29, 32–33
Communicative experience, 9–10, 23, 48
Consensual action, 1–2, 4, 97, 118–119, 208n
Consensual regulation of action conflicts, 78, 88, 91, 116, 119–120, 156, 161, 168; formal procedures in, 184, 205
Critical social theory, 69–71, 96

Defense mechanisms, theory of, 92–93